FUNDAMENTALS OF TOPOLOGY

BENJAMIN T. SIMS
Eastern Washington State College

MACMILLAN PUBLISHING CO., INC.
New York
COLLIER MACMILLAN PUBLISHE
London

Macmillan Publishing Co., Inc.
866 Third Avenue, New York, New York 10022

Collier Macmillan Canada, Ltd.

Library of Congress Cataloging in Publication Data

Sims, Benjamin T
Fundamentals of topology.

Includes index.
1. Topology. I. Title.
QA611.S495 514 74-33100
ISBN 0-02-410640-2

Printing: 1 2 3 4 5 6 7 8 Year: 6 7 8 9 0

FUNDAMENTALS OF TOPOLOGY

To the memory of my mother and father—

Whose sacrifices and encouragement made it possible

PREFACE

MY purpose in writing this book has been a desire to make available to undergraduates a unified treatment of the fundamental concepts of both point-set and algebraic topology. Such a book in the English language is practically nonexistent, although some introductory topology books do include a chapter on homotopy theory. At the graduate level, such unified treatments are provided by Hocking and Young's *Topology* and Schubert's *Topology*. Unfortunately, these are not at all well suited for use with undergraduates. I hope this book will provide both the undergraduate and the beginning graduate student with an outline ("overview") of the fundamentals of topology, encourage further study in this area, and enable them quickly to reach the frontiers of current mathematical research.

Let me say a few words initially concerning the contents of the book. Certain sections are starred, which indicates that they are optional. The remaining unstarred sections form what I shall refer to as the "basic core" of the text. Classroom testing has demonstrated that this basic core can be successfully taught to undergraduates in one semester or two quarters. In a one-quarter course, it is necessary to omit Chapter 8. With gifted undergraduate and beginning graduate students, one can cover in a semester (or two quarters) both the basic core and some of the optional topics such as uniform spaces, paracompactness, developable spaces, and so forth. If an entire year is available, the entire text can be thoroughly covered and even supplemented through the study of specific research papers chosen from the bibliography.

Next, I would like to emphasize the importance of the examples and exercises in this text. Acquaintance with a large number of examples and counterexamples is indispensable to the mathematician in general and the topologist in particular. Also, the exercises are to be viewed as an integral part of the text, inviting the student's participation in developing additional topological concepts and providing meaningful extensions of those developed in the text. In this way, the student has the creative experience of learning mathematics by doing mathematics, a method so elegantly espoused by the late Professor R. L. Moore.

The author is grateful for the many sources that have provided ideas for this particular treatment of topology, and they are listed in the bibliography. Special thanks are also due to Professor D. E. Sanderson and those students of mine who made numerous suggestions for the improvement of the manuscript, and to Mrs. Katherine Re, Mrs. Doris Foulon, and Mrs. Anne Biehl for their expert typing of the manuscript. The author also acknowledges the constant cooperation and assistance of the editors and staff of Macmillan

Publishing Co., Inc. Last, but not least, a bouquet of roses to my family, who have had to endure my late hours and frequent absences during the preparation of this manuscript.

B. T. S.

Cheney, Washington

CONTENTS

FUNDAMENTALS OF TOPOLOGY

CHAPTER 0
Introduction

0-1 Logic

In this preliminary chapter we discuss briefly a number of basic mathematical concepts that will be used in our subsequent development of the fundamentals of topology. These include sets, relations, functions, orderings, lattices, groups, homomorphisms, and the axiom of choice. In this section we summarize some of the rudiments of elementary logic.

A *proposition* is understood to be a meaningful sentence that is verifiable as being either true or false; e.g., "It's raining today" and "John is watching television" are propositions. Such sentences as "Open the window" and "x is a real number" are not propositions. We shall denote propositions by the letters $p, q, r, \ldots$. Through the use of *logical connectives*, we can form compound propositions from simple propositions p and q as follows:

(1) Negation: "Not p" or "p is false." $(\sim p)$
(2) Disjunction: "Either p is true or q is true." $(p \vee q)$
(3) Conjunction: "Both p and q are true." $(p \wedge q)$
(4) Implication: "If p is true, then q is true." $(p \rightarrow q)$
(5) Equivalence: "p is true if and only if q is true." $(p \leftrightarrow q$ or $p \equiv q)$

Henceforth the phrase "if and only if" will be denoted by "iff."

It is a useful exercise to construct the "truth tables" for the five compound propositions obtained above by use of the logical connectives. These are obtained by substituting the various possible combinations of the values T and F for p and q in each of the preceding compound propositions.

p	$\sim p$
T	F
F	T

p	q	$p \vee q$
T	T	T
T	F	T
F	T	T
F	F	F

p	q	$p \wedge q$
T	T	T
T	F	F
F	T	F
F	F	F

p	q	$p \rightarrow q$	$q \rightarrow p$	$p \leftrightarrow q\ [(p \rightarrow q) \wedge (q \rightarrow p)]$
T	T	T	T	T
T	F	F	T	F
F	T	T	F	F
F	F	T	T	T

By constructing their truth tables, the reader should verify that the implication $p \to q$ has the same truth values as its *contrapositive*, $\sim q \to \sim p$, and thus is logically equivalent to it. This provides the justification for the "indirect method" of proof in mathematics, whereby we show that the denial of the desired conclusion q implies a denial of our hypothesis p.

In mathematics, we frequently consider statements that are "universal" or "existential" in nature, such as "For all real numbers x and y, $(x + y)^2 = x^2 + 2xy + y^2$" and "There exist real numbers a and b such that $a^2 + b^2 = 25$." The quantifier "for all (every)" is denoted by the symbol $\forall$, and the quantifier "there exists" is denoted by the symbol $\exists$.

EXERCISES

0-1. Use the "truth-table method" to establish the following equivalences:

(a) $\sim\sim p \equiv p$.

(b) $(\sim p \vee q) \equiv (p \to q)$.

(c) $\sim(p \vee q) \equiv (\sim p \wedge \sim q)$.

(d) $\sim(p \wedge q) \equiv (\sim p \vee \sim q)$.

0-2. $q \to p$ is the *converse* and $\sim p \to \sim q$ is the *inverse* of the statement $p \to q$. Show that the converse and the inverse have the same truth value and thus are equivalent.

0-3. Use the results of Exercise 0–1 to show that $\sim(p \wedge \sim p) \equiv (\sim p \vee p)$. In Aristotelian logic, the statement on the left is referred to as the "law of noncontradiction," and the statement on the right is called the "law of the excluded middle."

0-4. A compound proposition that is true regardless of the truth values assigned to the simple propositions from which it is compounded is called a *tautology*. Show that $[(p \to q) \wedge (q \to r)] \to (p \to r)$ is a tautology, i.e., that implication is transitive.

0-5. Show that the following principles of valid inference are tautologies:

(a) $[(p \to q) \wedge p] \to q$ (Modus Ponens).

(b) $[(p \to q) \wedge \sim q] \to \sim p$ (Modus Tollens).

0-2 Sets, Relations, and Functions

By "set" we shall understand a collection S of elements called "points" whose membership (composition) is well defined. However, we do not formally define either "set" or "point." In any particular discussion, the set U of all points under consideration constitutes the *universal set* for that discussion. If p is an element of S, we use the notation "$p \in S$." If p is not an element of S, we use the notation "$p \notin S$." In defining the membership of a set S, we find useful the "set builder notation" $S = \{x \mid P(x)\}$; i.e., S is the set of all $x \in U$ such that $P(x)$ is true. As an example, if U is the set of integers and $S = \{x \mid x \text{ is divisible by } 2\}$, then S is the set of even integers.

DEFINITION 0-1. $\varnothing = \{x \mid x \neq x\}$ is called the *empty* (*null*) *set*.

DEFINITION 0-2
(1) A is a *subset* of B ($A \subset B$) iff each element of A is also an element of B.
(2) $A = B$ iff $A \subset B$ and $B \subset A$.
(3) A is a *proper subset* of B ($A \subsetneq B$) iff $A \subset B$ and $A \neq B$.

DEFINITION 0-3. Let A and B be subsets of U.
(1) The *union* of A and B is the set $A \cup B = \{x \mid x \in A \text{ or } x \in B\}$.
(2) The *intersection* of A and B is the set $A \cap B = \{x \mid x \in A \text{ and } x \in B\}$.
(3) The *Cartesian product* of A and B is the set $A \times B = \{\langle x, y\rangle \mid x \in A, y \in B\}$ of all ordered pairs $\langle x, y\rangle$.

The concepts described in Definitions 0-2 and 0-3 are illustrated in Figures 0-1 through 0-4 by the use of "Venn diagrams."

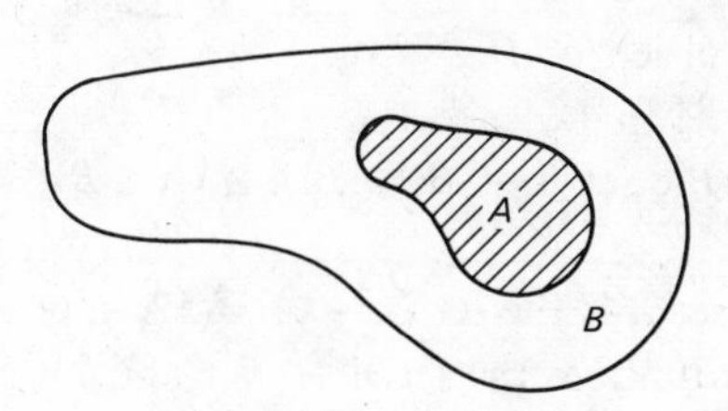

FIGURE 0-1. $A \subsetneq B$

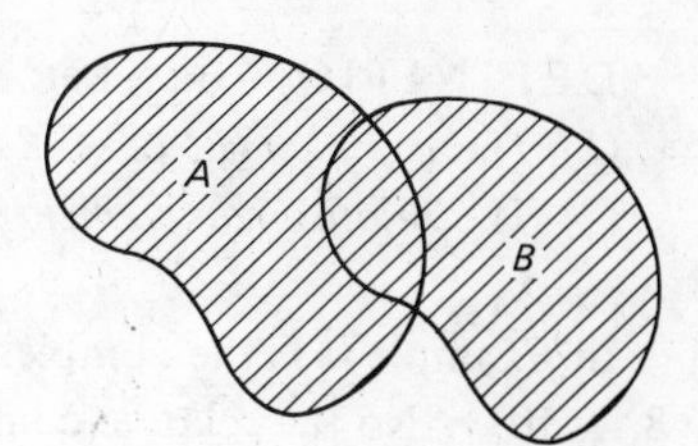

FIGURE 0-2. $A \cup B$

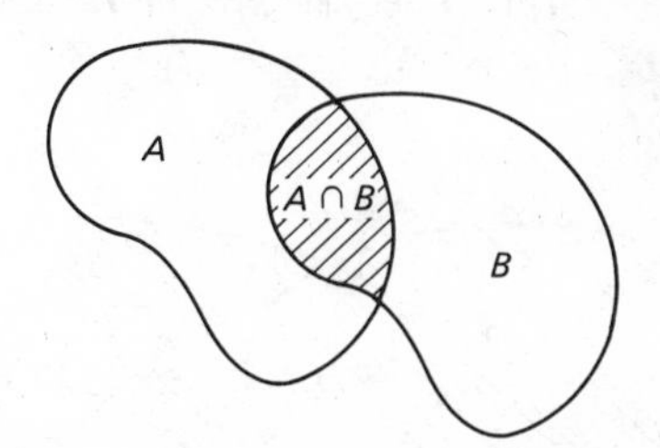

FIGURE 0-3. $A \cap B$

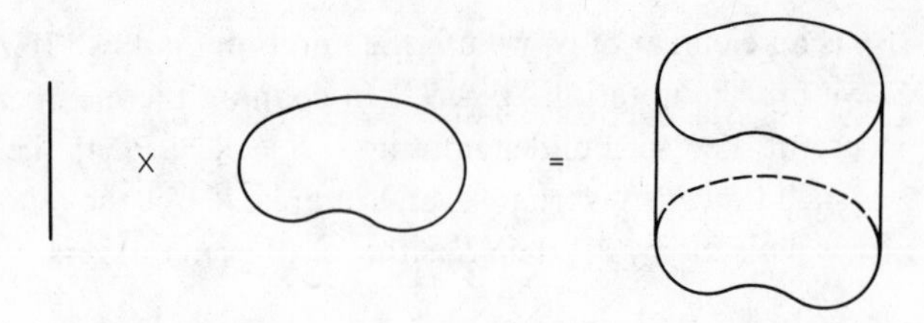

FIGURE 0-4

EXAMPLE 0-1. Let $U = \{1, 2, 3, 4, 5, 6, 7, 8, 9, 10\}$, $A = \{1, 3, 5, 7\}$, $B = \{3, 5, 9\}$, and $C = \{5, 7\}$. Then

(1) $C \subsetneq A$.
(2) $A \cup B = \{1, 3, 5, 7, 9\}$.
(3) $B \cup C = \{3, 5, 7, 9\}$.
(4) $A \cap B = \{3, 5\}$.
(5) $B \cap C = \{5\}$.
(6) $A \times B = \{\langle 1, 3\rangle, \langle 1, 5\rangle, \langle 1, 9\rangle, \langle 3, 3\rangle, \langle 3, 5\rangle, \langle 3, 9\rangle, \langle 5, 3\rangle, \langle 5, 5\rangle, \langle 5, 9\rangle, \langle 7, 3\rangle, \langle 7, 5\rangle, \langle 7, 9\rangle\}$.
(7) $B \times A = \{\langle 3, 1\rangle, \langle 3, 3\rangle, \langle 3, 5\rangle, \langle 3, 7\rangle, \langle 5, 1\rangle, \langle 5, 3\rangle, \langle 5, 5\rangle, \langle 5, 7\rangle, \langle 9, 1\rangle, \langle 9, 3\rangle, \langle 9, 5\rangle, \langle 9, 7\rangle\}$.

DEFINITION 0-4. If A and B are subsets of U, then A and B are *disjoint* iff $A \cap B = \varnothing$.

DEFINITION 0-5. Let A and B be subsets of U.
(1) The *complement* of A is the set $U - A = \{x \mid x \notin A\}$.
(2) The *relative complement* of B in A is the set $A - B = \{x \in A \mid x \notin B\}$.

In Example 0-1, the complement of the set A is the set $U - A = \{2, 4, 6, 8, 9, 10\}$. Also the relative complement of B in A is the set $A - B = \{1, 7\}$.

REMARK. Figure 0-5 shows that $A \cup B = (A - B) \cup (B - A) \cup (A \cap B)$. The set $(A - B) \cup (B - A)$ is called the *symmetric difference* and is denoted by $A \,\Delta\, B$. Thus $A \cup B = (A \,\Delta\, B) \cup (A \cap B)$.

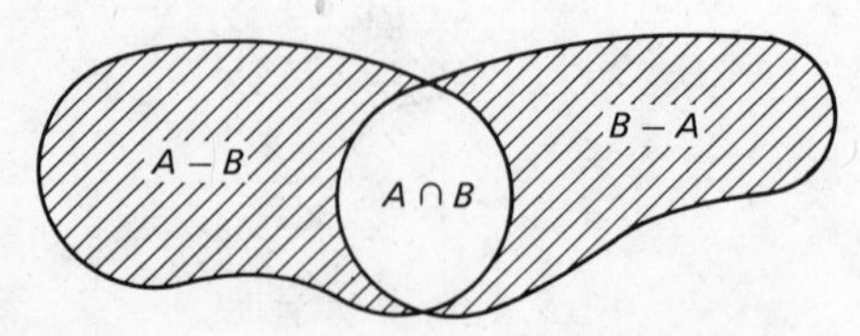

FIGURE 0-5. $A \,\Delta\, B = (A - B) \cup (B - A)$

Functions play an important role in topology as in analysis. Being single-valued relations, functions are special kinds of sets which we describe in our next two definitions.

DEFINITION 0-6. Let A and B be subsets of U.

(1) $R \subset A \times B$ is a *relation from A into B* iff $\forall a \in A\ \exists b \in B$ such that $\langle a, b \rangle \in R$. The *domain* of R is A, and the *range* of R is the set $\{b \in B \mid \langle a, b \rangle \in R$ for some $a \in A\}$. A relation from A into A is called a *relation on A*.

(2) $R \subset A \times B$ is a *relation from A onto B* iff R is a relation from A into B such that the range of R is B.

(3) If R is a relation from A into B, then the *inverse* of R is the set $R^{-1} = \{\langle b, a \rangle \mid \langle a, b \rangle \in R\}$.

EXAMPLE 0-2. Let I^+ denote the set of positive integers and $R = \{\langle m, n \rangle \mid m, n \in I^+$ and $n - m \in I^+\}$. Clearly, R is the relation on I^+ that orders the positive integers according to magnitude.

DEFINITION 0-7. Let A and B be subsets of U:

(1) $f: A \to B$ is a *function from A into B* iff f is a relation from A into B such that $\forall a \in A$ there exists a unique $b \in B$ with $\langle a, b \rangle \in f$.

(2) $f: A \to B$ is a *function from A onto B* iff f is a function from A into B such that the range $f(A)$ of f is B.

(3) If $f: A \to B$ is a function, then f is *injective* (1-1) iff $\langle a, b \rangle, \langle c, b \rangle \in f$ implies that $a = c$, and f is *bijective* iff f is injective and onto.

(4) If $f: A \to B$ is a function and $C \subset A$, then the *restriction* of f to C $(f \mid C)$ is defined by $(f \mid C)(x) = f(x)$, $\forall x \in C$, and the *inclusion function* $i: C \to A$ is defined by $i(x) = x$, $\forall x \in C$.

(5) If $f: A \to B$ is a function and $C \subset B$, then the set $f^{-1}(C) = \{x \in A \mid f(x) = y$ for some $y \in C\}$ is called the *preimage* of C under f.

EXAMPLE 0-3. Let $A = \{a, b, c\}$ and $B = \{1, 2, 3\}$. Observe that $R_1 = \{\langle a, 1 \rangle, \langle b, 1 \rangle, \langle c, 1 \rangle\}$ is a constant function, $R_2 = \{\langle a, 2 \rangle, \langle b, 3 \rangle, \langle c, 1 \rangle\}$ is a bijective function, and $R_3 = \{\langle a, 1 \rangle, \langle a, 2 \rangle, \langle b, 2 \rangle, \langle b, 3 \rangle, \langle c, 1 \rangle\}$ is a relation from A onto B that is not a function. The reader should examine additional subsets of $A \times B$ and determine which are functions (into, onto, injective, bijective), which are merely relations, and which are neither.

DEFINITION 0-8. If $f: A \to B$ and $g: B \to C$ are functions, then the *composition* $g \circ f: A \to C$ is the function given by $(g \circ f)(x) = g(f(x))$, $\forall x \in A$.

For a given collection of objects, we frequently find it desirable to "file" or "index" the objects in the collection. Our next definition formalizes this procedure.

DEFINITION 0-9. Let Λ be a nonempty set and φ the function defined by the equation $\varphi(\alpha) = S_\alpha \subset U$, $\forall \alpha \in \Lambda$. The collection $\{S_\alpha \mid \alpha \in \Lambda\}$ is said to be *indexed* by the *index set* Λ using the *index function* φ.

Functions that have the set I^+ of positive integers as their domain are called sequences.

DEFINITION 0-10

(1) A *sequence* in S is a function $x : I^+ \to S$, and we use the notation $\{x_n\}_{n \in I^+}$ rather than $\{x(n) \mid n \in I^+\}$ to denote the range of x.

(2) If $x : I^+ \to S$ is a sequence in S, a *subsequence* of x is a composition $x \circ y$, where $y : I^+ \to I^+$ is a function such that $y(i) > y(j)$ iff $i > j$.

EXAMPLE 0-4. Let $x : I^+ \to R$ (the reals) be given by $x_n = x(n) = 1/n$, $\forall n \in I^+$. Then x is a sequence that we denote by $\{1/n\}_{n \in I^+}$ rather than the strictly correct notation $\{\langle n, 1/n\rangle \mid n \in I^+\}$. If we let $y : I^+ \to I^+$ be the function given by $y(n) = 2n$, $\forall n \in I^+$, the composition $x \circ y : I^+ \to R$ is a subsequence of x that we denote by $\{1/2n\}_{n \in I^+}$.

We have defined a function as a single-valued relation from a set A into a set B. Another important type of relation on a set S is the "equivalence relation." An equivalence relation on S determines a partition of S into a collection of disjoint subsets called "equivalence classes."

DEFINITION 0-11. An *equivalence relation* R on a set $S \neq \varnothing$ is a relation on S that is

(1) Reflexive: $\langle x, x\rangle \in R$, $\forall x \in S$.

(2) Symmetric: $\langle x, y\rangle \in R$ implies that $\langle y, x\rangle \in R$, $\forall x, y \in S$.

(3) Transitive: $\langle x, y\rangle, \langle y, z\rangle \in R$ imply that $\langle x, z\rangle \in R$, $\forall x, y, z \in S$.

For each $x \in S$, the set $\{y \in S \mid \langle x, y\rangle \in R\}$ is the *R-equivalence class* of x and is denoted by $[x]$.

DEFINITION 0-12. A collection $\mathscr{P} = \{P_\alpha \mid \alpha \in \Lambda\}$ of subsets of a set S is a *partition* of S iff $S = \bigcup\{P_\alpha \mid \alpha \in \Lambda\}$ and, if $P_\alpha, P_\beta \in \mathscr{P}$, then $P_\alpha = P_\beta$ or $P_\alpha \cap P_\beta = \varnothing$.

THEOREM 0-1. *If R is an equivalence relation on S and $P_x = [x]$, $\forall x \in S$, then the collection $\mathscr{P} = \{P_x \mid x \in S\}$ is a partition of S.*

Proof. Since $\langle x, x\rangle \in R$, we have $x \in P_x$, $\forall x \in S$. Thus $\bigcup\{P_x \mid x \in S\} = S$. Moreover, if $z \in P_x \cap P_y$, then $\langle x, z\rangle$, $\langle y, z\rangle \in R$. Since R is symmetric, $\langle z, y\rangle \in R$ also. The transitivity of R implies that $\langle x, y\rangle \in R$, and $P_x = P_y$.

EXAMPLE 0-5. Let I be the set of integers and let $n \pmod p$ denote the remainder when $n \in I$ is divided by $p \in I^+$. For $m, n \in I$, let $\langle m, n\rangle \in R$ iff $m \pmod p = n \pmod p$. It is easily checked that R is an equivalence relation on I known as "congruence modulo p." R partitions I into the collection $\{[0], [1], \ldots, [p-1]\}$ of disjoint equivalence classes.

Although a development of the transfinite cardinal numbers and a discussion of the concept "power of a set" will not be given here, we frequently need to distinguish between finite and infinite sets and also between countable and uncountable sets. Our next definition provides us with the criteria for doing this.

DEFINITION 0-13

(1) S is *finite* iff $S = \varnothing$ or $\exists n \in I^+$ and a function $f: \{1, 2, \ldots, n\} \xrightarrow{\text{onto}} S$. Otherwise, S is *infinite*.
(2) S is *denumerable* (*countably infinite*) iff there is a 1-1 function $f: I^+ \xrightarrow{\text{onto}} S$. We denote by $\aleph_0$ the cardinal number of any denumerable set.
(3) S is *countable* iff S is finite or denumerable. Otherwise, S is uncountable.

EXAMPLE 0-6. The set I^+ is denumerable, since the identity function $i: I^+ \to I^+$, given by $i(n) = n$, $\forall n \in I^+$, is 1-1 and onto. The set R of real numbers is uncountable, however. Indeed, the subset $S = \{x \in R \mid 0 < x \leq 1\}$ is uncountable. For, if S were denumerable, we could denote it thus: $S = \{x_1, x_2, \ldots, x_n, \ldots\}$, where each x_n has a unique (nonterminating) decimal expansion $0.a_{n1}a_{n2}a_{n3} \cdots a_{nn} \cdots$. For example, we use for $\frac{1}{4}$ the nonterminating decimal expansion $0.24999\ldots$ rather than the terminating decimal expansion $0.25000\ldots$. Consider the nonterminating decimal $0.a_{11}a_{22}\cdots a_{nn}\cdots$ and form a decimal $y = 0.b_1b_2\ldots b_n\ldots$ with $b_n \neq 0$ and $b_n \neq a_{nn}$, $\forall n \in I^+$. As a result of the absence of zeros and our manner of construction, y has a nonterminating decimal representation distinct from those representing the elements of S. Thus our assumption that S was denumerable must be false. We denote by c the cardinal number of any set whose elements can be put into 1-1 correspondence with those of S. Thus the cardinal number of R is c. Moreover, the famous "continuum hypothesis" asserts that there is no cardinal number between $\aleph_0$ and c.

EXERCISES

0-6. Let $\{C_\alpha \mid \alpha \in \Lambda\}$ be a collection of subsets of S. Verify the following equivalences, known as DeMorgan's laws:

(a) $S - \bigcup\{C_\alpha \mid \alpha \in \Lambda\} = \bigcap\{S - C_\alpha \mid \alpha \in \Lambda\}$.
(b) $S - \bigcap\{C_\alpha \mid \alpha \in \Lambda\} = \bigcup\{S - C_\alpha \mid \alpha \in \Lambda\}$.

0-7. Let $f: A \to B$ be a function and $\{A_\alpha \mid \alpha \in \Lambda\}$ a collection of subsets of A. Show that

(a) $f(\bigcup\{A_\alpha \mid \alpha \in \Lambda\}) = \bigcup\{f(A_\alpha) \mid \alpha \in \Lambda\}$.
(b) $f(\bigcap\{A_\alpha \mid \alpha \in \Lambda\}) \subset \bigcap\{f(A_\alpha) \mid \alpha \in \Lambda\}$.

0-8. Let I^+ be the set of positive integers and R a relation on I^+ with the following properties:

(a) If n is even and m is odd, then $\langle n, m\rangle \notin R$.
(b) If n and m are both even or both odd, then $\langle n, m\rangle \in R$ iff there exists an integer k (not necessarily positive) such that $n = m + k$.

Is R an equivalence relation on I^+? If so, describe the equivalence classes.

0-9. Let $\mathscr{P} = \{P_\alpha \mid \alpha \in \Lambda\}$ be a partition of a set S. For all $x, y \in S$, let $\langle x, y\rangle \in R$ iff $x, y \in P_\alpha$ for some $\alpha \in \Lambda$. Show that R is an equivalence relation on S and describe the equivalence classes.

0-10. Let $f: A \xrightarrow{\text{onto}} B$ be a function. Show that if A is finite, then B is finite, and if A is countable, then B is countable.

0-11. $R^\# = \{\langle p, q\rangle \mid p, q \in I, q \neq 0\}$ is the set of rational numbers with the understanding that $\langle p_1, q_1\rangle = \langle p_2, q_2\rangle$ iff $p_1 q_2 = p_2 q_1$. $R^\#$ is infinite, since it contains the infinite subset $I^+ = \{\langle p, 1\rangle \mid p \in I^+\}$. Show that $R^\#$ is countable, however, by showing that it is denumerable.

0-3 Orderings, Lattices, and Chains

In this section we introduce the notions of a partial ordering, a linear ordering, and a well-ordering in terms of which two types of algebraic structures known as lattices and chains are defined. A Boolean algebra is then definable as a lattice that is both complemented and distributive.

DEFINITION 0-14. A *partial ordering* $\prec$ on $S \neq \varnothing$ is a relation on S that is

(1) Reflexive: $x \prec x, \forall x \in S$.

(2) Antisymmetric: $x \prec y$ and $y \prec x$ implies that $x = y$, $\forall x, y \in S$.
(3) Transitive: $x \prec y$ and $y \prec z$ implies that $x \prec z$, $\forall x, y, z \in S$.
The term "$x \prec y$" is read "x precedes y." Furthermore, we shall write "$x \succ y$" (i.e., "x follows y") iff $y \prec x$.

Next, we define the concepts of maximal and minimal elements and least and greatest element in a partially ordered set.

DEFINITION 0-15. Let $S \neq \varnothing$ be partially ordered by $\prec$:
(1) m is a *maximal element* of S iff $\nexists x \in S - \{m\}$ such that $m \prec x$.
(2) m is a *minimal element* of S iff $\nexists x \in S - \{m\}$ such that $m \succ x$.
(3) m is a *greatest element* of S iff $m \succ x$, $\forall x \in S$.
(4) m is a *least element* of S iff $m \prec x$, $\forall x \in S$.

REMARK. Observe that a greatest element is a maximal element and a least element is a minimal element. Also, whenever they exist, a greatest element or a least element is necessarily unique.

EXAMPLE 0-7. Let L be the set of all functions mapping $[0, 1]$ into $[0, 1]$. If $f, g \in L$, then $f \prec g$ iff $f(x) \leq g(x)$, $\forall x \in [0, 1]$. The function $f(x) \equiv 0$ is the least element of L and the function $f(x) \equiv 1$ is the greatest element. Moreover, they are the only maximal and minimal elements of L.

Certain partially ordered sets $\langle L, \prec \rangle$ are structures that we call "lattices." We preface our definition of "lattice" with the usual definitions of upper and lower bounds and least upper and greatest lower bounds.

DEFINITION 0-16. Let $\langle L, \prec \rangle$ be a partially ordered set and $\varnothing \neq A \subset L$.
(1) $b \in L$ is an *upper bound* for A iff $a \prec b$, $\forall a \in A$.
(2) $b \in L$ is a *lower bound* for A iff $b \prec a$, $\forall a \in A$.
(3) An upper bound b for A is the *least upper bound* ("sup") of A iff $b \prec c$ for every upper bound c for A.
(4) A lower bound b for A is the *greatest lower bound* ("inf") of A iff $c \prec b$ for every lower bound c for A.

DEFINITION 0-17. A partially ordered set $\langle L, \prec \rangle$ is a *lattice* iff $\forall a, b \in L$, $\{a, b\}$ has a greatest lower bound $a \cdot b \in L$ and a least upper bound $a + b \in L$. The partially ordered set in Figure 0-6(b) is a lattice, and the partially ordered set in Figure 0-6(a) is not a lattice.

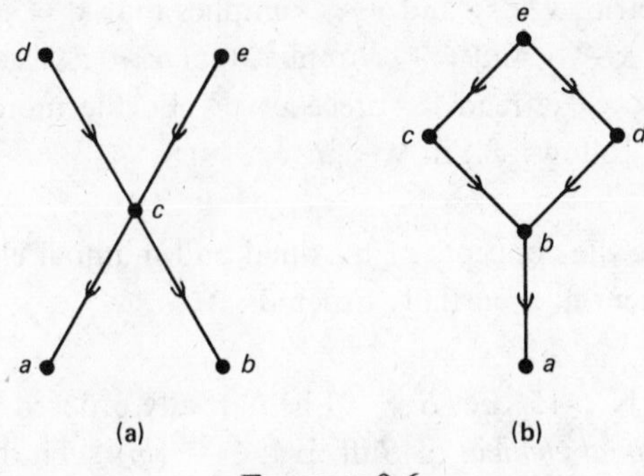

FIGURE 0-6

EXAMPLE 0-8. Let L be the set of positive integers. If $a, b \in L$, then $a \prec b$ iff a is a divisor of b. For all $a, b \in L$, $a \cdot b = \inf \{a, b\}$ = greatest common divisor of a and b, and $a + b = \sup \{a, b\}$ = least common multiple of a and b. $\langle L, \prec \rangle$ is a lattice with least element 1 and no greatest element.

Next we introduce the notion of a "total ordering" and use it to define a second algebraic structure known as a "chain." Also, the concept of a "well-ordering" is introduced here and will be used in the next section.

DEFINITION 0-18. A partial ordering $\prec$ on S is a *total* (*complete*, *simple*) *ordering* iff $\forall a, b \in S$ either $a \prec b$ or $b \prec a$.

DEFINITION 0-19. Let $\langle S, \prec \rangle$ be a partially ordered set and $\varnothing \neq A \subset S$. $\langle A, \prec | A \rangle$ is a *chain* in $\langle S, \prec \rangle$ iff $\prec | A$ is a total ordering on A.

REMARK. In Example 0-8 the subset $A = \{2^n \mid n \in I^+\}$ is totally ordered by $\prec | A$. Thus $\langle A, \prec | A \rangle$ is a chain in $\langle L, \prec \rangle$.

DEFINITION 0-20. A total ordering $\prec$ on S is a *well-ordering* for S iff each nonempty subset A of S has a least element relative to $\prec | A$. (For example, the set I^+ of positive integers is well ordered by the order relation "less than" described in Example 0-2.)

We conclude this section with definitions of a complemented lattice and a distributive lattice. A *Boolean algebra* is defined to be a lattice that is both complemented and distributive.

DEFINITION 0-21. Let $\langle L, \prec \rangle$ be a lattice with least element 0 and greatest element 1. $\langle L, \prec \rangle$ is *complemented* iff $\forall a \in L \; \exists x \in L$ such that $a + x = 1$ and $a \cdot x = 0$. Such an x is called the *complement* of a.

DEFINITION 0-22. A lattice $\langle L, \prec \rangle$ is *distributive* iff $(a + b) \cdot c = (a \cdot c) + (b \cdot c)$, $\forall a, b, c \in L$.

REMARK. The lattice $\langle L, \prec \rangle$ of Example 0-8 is distributive but not complemented, since it has no greatest element. Also, if we let $f \cdot g$ be the function given by $(f \cdot g)(x) = \min \{f(x), g(x)\}$ and $f + g$ the function given by $(f + g)(x) = \max \{f(x), g(x)\}$, $\forall x \in [0, 1]$, then $\langle L, \prec \rangle$ in Example 0-7 is a distributive lattice with both a least and a greatest element. However, the reader may easily verify that it is not complemented either.

EXERCISES

0-12. If $\langle L, \prec \rangle$ is a partially ordered set and $\varnothing \neq A \subset L$, show that that $\langle A, \prec \mid A \rangle$ is a partially ordered set.

0-13. If $\langle L, \prec \rangle$ is a lattice and A is a finite subset of L, show that A has both a least upper bound and a greatest lower bound in L.

0-14. Let $S \neq \varnothing$ and $\langle L, \prec \rangle$ be the collection of all subsets of S partially ordered by set inclusion. If $A, B \in L$, then $A \cdot B = A \cap B$ and $A + B = A \cup B$. Show that $\langle L, \prec \rangle$ is a Boolean algebra.

0-15. Let L be the class of all decidable propositions, where $p \prec q$ iff $p \to q$, $\forall p, q \in L$. Moreover, $p \cdot q = p \wedge q$ and $p + q = p \vee q$, $\forall p, q \in L$. Show that $\langle L, \prec \rangle$ is a distributive lattice. (Actually, it is isomorphic to a Boolean algebra of two elements, $\{0, 1\}$, hence the name "two-valued logic.")

0-4 Equivalents of the Axiom of Choice

In 1963, P. J. Cohen showed that both the "Axiom of Choice" and the "Continuum Hypothesis" are independent of the axioms of Zermelo–Fraenkel set theory. Thus we agree to add both of these as axioms, since the proofs of some of the results in topology utilize them. Indeed, Tychonoff's theorem, which states that the product of any number of compact topological spaces is compact, is equivalent to the axiom of choice. Two additional equivalents of the Axiom of Choice, Zorn's lemma and the Well-Ordering Principle, are stated below. The Well-Ordering Principle is used in our construction of the ordinals.

AXIOM OF CHOICE. *If* $\{S_\alpha \mid \alpha \in \Lambda\}$ *is a nonempty collection of nonempty sets, there is a function* $c : \Lambda \to \bigcup\{S_\alpha \mid \alpha \in \Lambda\}$ *such that* $c(\alpha) \in S_\alpha$, $\forall \alpha \in \Lambda$.

ZORN'S LEMMA. *If* $\langle S, \prec \rangle$ *is a partially ordered set such that each chain in* $\langle S, \prec \rangle$ *has an upper bound in S, then S has a maximal element.*

WELL-ORDERING PRINCIPLE. *Every set can be well ordered.*

THEOREM 0-2. *The Axiom of Choice, Zorn's lemma, and the Well-Ordering Principle are equivalent statements.*

Proof. The proof that the Axiom of Choice implies Zorn's lemma is rather long and will not be given here. It is found in Hall and Spencer's *Elementary Topology* and Halmos's *Naïve Set Theory*, among many other places. The proof that the Well-Ordering Principle implies the Axiom of Choice is left as an easy exercise for the reader. We sketch next the proof that Zorn's lemma implies the Well-Ordering Principle.

Let $\varnothing \neq A \subset S$ and $W(A)$ be the set of all well-orderings on A. If we let $\mathscr{A} = \{\langle A, w(A)\rangle \mid A \subset S, w(A) \in W(A)\}$, we can construct a partial ordering $\prec$ on $\mathscr{A}$ as follows: $\langle A, w(A)\rangle, \langle B, w(B)\rangle \in \mathscr{A}$ implies that $\langle A, w(A)\rangle \prec \langle B, w(B)\rangle$ iff $A \subset B$, $w(A) = w(B) \mid A$, and $x \in A$ and $y \in B - A$ imply that $\langle x, y\rangle \in w(B)$. $\mathscr{A} \neq 0$, since any total ordering of a finite subset A of S is a well-ordering on A. If $\mathscr{C} = \{\langle A_\alpha, w(A_\alpha)\rangle \mid \alpha \in \Lambda\}$ is any chain in $\langle \mathscr{A}, \prec\rangle$, let $C = \bigcup\{A_\alpha \mid \alpha \in \Lambda\}$ and construct a partial ordering $\prec_c$ on C as follows: $x, y \in C$ implies that $\exists \beta, \gamma \in \Lambda$ such that $x \in A_\beta$, $y \in A_\gamma$, where $A_\beta \subset A_\gamma$ and $w(A_\beta) = w(A_\gamma) \mid A_\beta$, and we let $x \prec_c y$ iff $\langle x, y\rangle \in w(A_\gamma)$. Also, $\prec_c$ is a well-ordering for C, since $\varnothing \neq D \subset C$ implies that $D \cap A_\alpha \neq \varnothing$ for some α and the least element of $D \cap A_\alpha$ is the least element of D. Since $\langle A_\alpha, w(A_\alpha)\rangle \prec \langle C, \prec_c\rangle$ for each $\alpha \in \Lambda$, $\langle C, \prec_c\rangle$ is an upper bound on $\mathscr{C}$ and Zorn's lemma implies that $\mathscr{A}$ has a maximal element $\langle M, w(M)\rangle$. Suppose that $\exists x \in S - M$ and let $N = M \cup \{x\}$. Construct a well-ordering $w(N)$ on N as follows: $\langle y, x\rangle \in w(N)$, $\forall y \in M$, and $w(N) \mid M = w(M)$. Thus $\langle N, w(N)\rangle \in \mathscr{A}$ and $\langle M, w(M) \prec \langle N, w(N)\rangle$. This contradicts the maximality of $\langle M, w(M)\rangle$. Thus $M = S$.

REMARK. Although the Well-Ordering Principle guarantees the *existence* of a well-ordering for every set, it should be pointed out that no one has yet *constructed* a well-ordering for the reals.

In Example 0-9 we utilize the Well-Ordering Principle to construct the set $\mathcal{O}$ of countable ordinals and the set $\mathcal{O}^* = \mathcal{O} \cup \{\Omega\}$, where Ω is the first uncountable ordinal.

EXAMPLE 0-9. The Well-Ordering Principle guarantees the existence of a well-ordering $\prec$ on $R - I^+$. Let $\alpha \notin R$ be arbitrary but fixed. We extend $\prec$ to a well-ordering on $R \cup \{\alpha\}$ as follows:

(1) $1 \prec 2 \prec \cdots \prec n \prec \cdots, \forall n \in I^+$.

(2) For each $n \in I^+$, $n \prec x$, $\forall x \in R - I^+$.

(3) $x \prec \alpha$, $\forall x \in R$.

Condition (3) implies that α has uncountably many predecessors. Let Ω be the least element of $R \cup \{\alpha\}$ with uncountably many predecessors. Denote by $\mathcal{O}$ the set of predecessors of Ω and by $\mathcal{O}^*$ the set $\mathcal{O} \cup \{\Omega\}$. Also, let ω be the least element of $\mathcal{O}$ that follows every $n \in I^+$. Thus each element of $\mathcal{O}$ has only countably many predecessors, ω is the first infinite ordinal, and Ω is the first uncountable ordinal.

We conclude this section with a result on ordinals that will be used later.

THEOREM 0-3. *If $A \subset \mathcal{O}$ is countable, then* $(\sup A) \in \mathcal{O}$.

Proof. Since Ω is an upper bound of A, then A has a least upper bound b. Let $B = \{x \in \mathcal{O} \mid x \in A$ or x precedes some element of $A\}$. Then B is countable, since A is countable and each of its elements has only countably many predecessors. Also, b is an upper bound for B, and each predecessor of b is in B. Since B is countable, $b \prec \Omega$ and $b \in \mathcal{O}$.

EXERCISES

0-16. Prove that the Well-Ordering Principle implies the Axiom of Choice.

0-17. Show that a set S is infinite iff there exists a sequence $\{p_n\}_{n \in I^+}$ of distinct points of S.

★0-5 Groups, Homomorphisms, and Isomorphisms

In this final section of our introduction, we discuss several algebraic concepts that will be used in Chapters 7 and 8, where we give a brief introduction to algebraic topology. These include group, subgroup, homomorphism, isomorphism, and quotient group.

DEFINITION 0-23. $\langle G, \circ \rangle$ is a *group* iff G is a nonempty set and $\circ : G \times G \to G$ is a function such that

(1) $\circ$ is associative: $a \circ (b \circ c) = (a \circ b) \circ c$, $\forall a, b, c \in G$.

(2) G contains an identity element with respect to $\circ$: $\exists e \in G$ such that $a \circ e = e \circ a = a$, $\forall a \in G$.

(3) Each $a \in G$ has an inverse element in G with respect to $\circ$: $\forall a \in G\ \exists a^{-1} \in G$ such that $a \circ a^{-1} = a^{-1} \circ a = e$.

DEFINITION 0-24. A group $\langle G, \circ \rangle$ is *Abelian* iff $a \circ b = b \circ a$, $\forall a, b \in G$.

DEFINITION 0-25. If $\langle G, \circ \rangle$ is a group, $\varnothing \neq H \subset G$, and $* = \circ \mid H \times H$, then $\langle H, * \rangle$ is a *subgroup* of $\langle G, \circ \rangle$ iff $\langle H, * \rangle$ is a group. Also, $\langle H, * \rangle$ is a *proper subgroup* of $\langle G, \circ \rangle$ iff $\langle H, * \rangle$ is a subgroup of $\langle G, \circ \rangle$ and $H \neq \{e\}, G$.

THEOREM 0-4. *If* $\langle G, \circ \rangle$ *is a group,* $\varnothing \neq H \subset G$ *and* $* = \circ \mid H \times H$, *then* $\langle H, * \rangle$ *is a subgroup of* $\langle G, \circ \rangle$ *iff* $a * b^{-1} \in H$, $\forall a, b \in H$.

Proof. The necessity of the condition is trivial from Definition 0-23. Suppose now that $a * b^{-1} \in H$, $\forall a, b \in H$. We show that $\langle H, * \rangle$ is a group and hence a subgroup of $\langle G, \circ \rangle$. Clearly, $*$ is associative, since $\circ$ is associative. Since $H \neq \varnothing$, $\exists a \in H$. Hence $a * a^{-1} = e \in H$ by our hypothesis. Again, by our hypothesis, we get $e * a^{-1} = a^{-1} \in H$, $\forall a \in H$. Since $\langle H, * \rangle$ satisfies the conditions of Definition 0-23, it is a group.

EXAMPLE 0-10. If $G = I$ (the set of integers) and $\circ = +$ (ordinary addition), then $\langle I, + \rangle$ is an Abelian group. Note that $e = 0$ and $n^{-1} = -n$, $\forall n \in I$. If $H = E$ (the set of even integers), then $\langle E, + \rangle$ is a proper subgroup of $\langle I, + \rangle$.

EXAMPLE 0-11. If G is the set of positive rationals and $\circ = \cdot$ (ordinary multiplication), then $\langle G, \cdot \rangle$ is an Abelian group with $e = 1$ and $(p/q)^{-1} = q/p$ for each rational p/q.

DEFINITION 0-26

(1) If $\langle G, \circ \rangle$ and $\langle H, * \rangle$ are groups and $f: G \to H$ is a function, then f is a *homomorphism* iff $f(a \circ b) = f(a) * f(b)$, $\forall a, b \in G$.

(2) If $\langle G, \circ \rangle$ and $\langle H, * \rangle$ are groups, e^* is the identity of $\langle H, * \rangle$ and $f: G \to H$ is a homomorphism, then $\langle f(G), * \mid f(G) \times f(G) \rangle$ is a subgroup of $\langle H, * \rangle$ called the *image* of $\langle G, \circ \rangle$ under f and $\langle f^{-1}(e^*), \circ \mid f^{-1}(e^*) \times f^{-1}(e^*) \rangle$ is a subgroup of $\langle G, \circ \rangle$ called the *kernel* of $\langle G, \circ \rangle$ under f.

(3) If $\langle G, \circ \rangle$ and $\langle H, * \rangle$ are groups, then $\langle G, \circ \rangle$ and $\langle H, * \rangle$ are *isomorphic* (i.e., $\langle G, \circ \rangle \cong \langle H, * \rangle$) iff there exists a 1-1 homomorphism $f: G \xrightarrow{\text{onto}} H$. Such a homomorphism is called an *isomorphism*.

REMARK. It is an easy exercise to show that a necessary and sufficient condition that a homomorphism $f: G \xrightarrow{\text{onto}} H$ be an isomorphism is that the kernel of G under f (henceforth abbreviated as "kerf") be $\{e\}$, where e is the identity element of $\langle G, \circ \rangle$.

DEFINITION 0-27

(1) If $\langle G, \circ \rangle$ is a group and $\varnothing \neq K \subset G$ such that each element of G can be expressed as a "$\circ$-product" of elements of K and their inverses, then K is a set of *generators* for $\langle G, \circ \rangle$. A group with a single generator is known as a *cyclic group.* (For example, the set of integers under addition is a cyclic group with one as generator.)

(2) If K is a set of generators for an Abelian group $\langle G, \circ \rangle$ such that the only "$\circ$-products" of elements of K and their inverses equal to the identity are of the form $x \circ x^{-1}$, then $\langle G, \circ \rangle$ is a *free Abelian group.* [For example, if Z denotes the additive Abelian group of integers, then the direct sum (Definition 0-28) $Z \oplus Z$ is a free Abelian group with two generators.]

REMARK. If K is a set of generators for a group $\langle G, \circ \rangle$ and $\langle H, * \rangle$ is any other group, then a homomorphism $f: G \to H$ is completely specified by the values of f on K. Thus, if $\langle G, \circ \rangle$ is a free Abelian group and $\langle H, * \rangle$ any Abelian group, and f assigns arbitrary (but fixed) values in H to the generators of $\langle G, \circ \rangle$, then f can be extended to a homomorphism of G into H.

DEFINITION 0-28. If $\langle G, + \rangle$ is an additive Abelian group and $\langle H_i, + \mid H_i \times H_i \rangle$ is a subgroup of $\langle G, + \rangle$, $\forall i \in \Lambda$ (countable) such that each element of G is uniquely expressible as a sum of elements, one from each H_i with only a finite number different from zero (the identity of $\langle G, + \rangle$), then G is called the *direct sum* of the H_i and is written $G = H_1 \oplus H_2 \oplus \cdots$.

We conclude our brief discussion of algebraic concepts with the definition of a quotient group and a statement of the fundamental isomorphism theorem for groups.

DEFINITION 0-29. Let $\langle G, + \rangle$ be an additive Abelian group and $\langle H, + \mid H \times H \rangle$ a subgroup of $\langle G, + \rangle$. G can be decomposed into a family $\{a + H \mid a \in G\}$ of subsets called *cosets,* where $a + H = \{a + b \mid b \in H\}$ for each $a \in G$. If $a + H$ and $b + H$ are cosets of G, then $(a + H) \oplus (b + H) = (a + b) + H$. Under coset addition, the family of cosets of H in G forms a group G/H called the *quotient group* of G with respect to H.

THEOREM 0-5. *If* $\langle G, +\rangle$ *and* $\langle G', *\rangle$ *are additive Abelian groups and* $f: G \xrightarrow{onto} G'$ *is a homomorphism with* $\ker f = H$, *then* $G/H \cong G'$.

Proof. Let $g : G/H \to G'$ be given by $g(a + H) = f(a)$, $\forall a \in G$. Clearly, g is onto, since f is onto by hypothesis. Also, $g[(a + H) \oplus (b + H)] = g[(a + b) + H] = f(a + b)$ and $g(a + H) = f(a)$ and $g(b + H) = f(b)$. Moreover, $f(a + b) = f(a) * f(b)$, since f is a homomorphism. Hence g is a homomorphism onto. Moreover, $\ker g = \{a + H \mid g(a + H) = f(a) = e'\} = \ker f = H = e + H$ (the identity coset). Thus, by a previous remark, g is an isomorphism.

EXERCISES

0-18. If $G = \{0, 1, 2, 3, 4, 5\}$ and $\oplus$ is addition modulo 6, show that $\langle G, \oplus\rangle$ is a group and determine all proper subgroups.

0-19. Show that a necessary and sufficient condition that a homomorphism $f: \langle G, \circ\rangle \xrightarrow{\text{onto}} \langle H, *\rangle$ be an isomorphism is that $\ker f = \{e\}$, where e is the identity of $\langle G, \circ\rangle$.

0-20. Show that if G is an additive Abelian group and H is a subgroup of G, then $x, y \in (a + H)$ for some $a \in G$ iff $x - y \in H$. Use this result to show that $\{a + H \mid a \in G\}$ is a partition of G and thus determines an equivalence relation on G.

0-21. Supply the details necessary to show that G/H is a group under coset addition (see Definition 0-29).

CHAPTER 1
Topological Spaces

1-1 Topologies

A topologist has often been facetiously described as a person who is unable to distinguish between a doughnut and a coffee cup, since the two objects are topologically equivalent. In a much more esoteric (but less intuitive) vein, we can describe topology as the study of topological properties of topological spaces. However, in order for this characterization of topology to be meaningful, it is necessary that we define the two concepts "topological space" and "topological property." In this chapter we do this and more. We indicate a variety of ways in which a set may be "topologized" (i.e., assigned a mathematical structure known as a "topology"). Topological properties and their inheritance by subspaces and product spaces are investigated. In addition, we define and study briefly several special classes of topological spaces, including distance spaces, uniform spaces, proximity spaces, and quotient spaces.

In this section we describe several ways to "topologize" a nonempty set S. We show that the collection $\mathscr{L}$ of all topologies on a set S is partially ordered by set inclusion with least element $\{\varnothing, S\}$ and greatest element $2^S = \{G \mid G \subset S\}$ (the "power set" of S). We leave to the reader the exercise of showing that $\mathscr{L}$ is actually a lattice when lattice sum and product are appropriately defined.

DEFINITION 1-1. Let $S \neq \varnothing$ and $\tau \subset 2^S$ (the collection of all subsets of S) such that the following three axioms hold:

(O1) $\varnothing \in \tau$ and $S \in \tau$.

(O2) If $G_\alpha \in \tau$, $\forall \alpha \in \Lambda$, then $\bigcup\{G_\alpha \mid \alpha \in \Lambda\} \in \tau$.

(O3) If $G_i \in \tau$ $(i = 1, 2, \ldots, n)$, then $\bigcap_{i=1}^n G_i \in \tau$.

Then τ is a *topology* on S, the elements of τ are called the "open sets," and the ordered pair $\langle S, \tau\rangle$ is called a *topological space.*

EXAMPLE 1-1. Let $S = \{a, b, c\}$. The reader can easily check that $\tau_1 = \{\varnothing, S\}$, $\tau_2 = \{\varnothing, \{a\}, S\}$, $\tau_3 = \{\varnothing, \{a\}, \{b, c\}, S\}$, $\tau_4 = \{\varnothing, \{a\}, \{a, b\}, \{a, c\}, S\}$, and $\tau_5 = \{\varnothing, \{a\}, \{b\}, \{c\}, \{a, b\}, \{a, c\}, \{b, c\}, S\} = 2^S$ satisfy axioms (O1)–(O3) and hence are topologies on S. We leave to the reader the exercise of finding all remaining subcollections of 2^S that are topologies on S. There are 29 topologies definable on S.

DEFINITION 1-2. Let $S \neq \varnothing$. The smallest topology on S (i.e., the one with the fewest elements) is $\{\varnothing, S\}$ and is called the *indiscrete topology*. The largest topology on S (i.e., the one with the most elements) is 2^S and is called the *discrete topology*. In Example 1-1, τ_1 is the indiscrete topology and τ_5 is the discrete topology for $S = \{a, b, c\}$.

THEOREM 1-1. *Let $\mathscr{L}$ be the collection of all topologies on $S \neq \varnothing$ and $\subset$ be set inclusion. Then $\langle \mathscr{L}, \subset \rangle$ is a partially ordered set with least element $\{\varnothing, S\}$ and greatest element 2^S.*

Proof. Since every topology on S is a subcollection of itself, $\subset$ is reflexive. Also, $\subset$ is antisymmetric, since $\tau_1 \subset \tau_2$ and $\tau_2 \subset \tau_1$ iff $\tau_1 = \tau_2$, where $\tau_1, \tau_2 \in \mathscr{L}$. Moreover, $\subset$ is transitive, since $\tau_1 \subset \tau_2$ and $\tau_2 \subset \tau_3$ imply that $\tau_1 \subset \tau_3$ for $\tau_1, \tau_2, \tau_3 \in \mathscr{L}$. Finally, $\{\varnothing, S\} \subset \tau \subset 2^S, \forall \tau \in \mathscr{L}$ from Definition 1-1.

DEFINITION 1-3. Let $\langle S, \tau \rangle$ be a topological space and $A \subset S$. Then, A is *closed* iff $S - A \in \tau$.

THEOREM 1-2. *Let $\langle S, \tau \rangle$ be a topological space.*

(C1) *$\varnothing$ and S are closed.*

(C2) *If $A_\alpha \subset S$ is closed $\forall \alpha \in \Lambda$, then $\bigcap\{A_\alpha \mid \alpha \in \Lambda\}$ is closed.*

(C3) *If $A_i \subset S$ is closed $(i = 1, 2, \ldots, n)$, then $\bigcup_{i=1}^n A_i$ is closed.*

Proof

(C1) $\varnothing$ and S are closed, since their respective complements S and $\varnothing$ are open (i.e., elements of τ).

(C2) Let $A_\alpha \subset S$ be closed, $\forall \alpha \in \Lambda$. This implies that $S - A_\alpha \in \tau, \forall \alpha \in \Lambda$. Also, $S - \bigcap\{A_\alpha \mid \alpha \in \Lambda\} = \bigcup\{S - A_\alpha \mid \alpha \in \Lambda\}$ by DeMorgan's law [see Exercise 0-6(b)]. Since $S - A_\alpha \in \tau, \forall \alpha \in \Lambda$, $\bigcup\{S - A_\alpha \mid \alpha \in \Lambda\} \in \tau$ by axiom (O2). Thus $\bigcap\{A_\alpha \mid \alpha \in \Lambda\}$ is closed by Definition 1-3.

(C3) Let $A_i \subset S$ be closed $(i = 1, 2, \ldots, n)$. This implies that $S - A_i \in \tau$ $(i = 1, 2, \ldots, n)$. Also, $S - \bigcup_{i=1}^n = \bigcap_{i=1}^n (S - A_i)$ by DeMorgan's law [see Exercise 0-6(a)]. Since $S - A_i \in \tau$ $(i = 1, 2, \ldots, n)$, $\bigcap_{i=1}^n (S - A_i) \in \tau$ by axiom (O3). Thus $\bigcup_{i=1}^n A_i$ is closed by Definition 1-3.

REMARK. Theorem 1-2 provides us with a second way of defining a topology on a nonempty set S. We begin with a collection $\mathscr{F} \subset 2^S$, which satisfies (C1)–(C3) of Theorem 1-2 and whose elements are called the "closed sets." Each such collection $\mathscr{F}$ determines a unique topology $\tau = \{G \subset S \mid S - G \in \mathscr{F}\}$ on S.

DEFINITION 1-4. Let $\langle S, \tau \rangle$ be a topological space and $A \subset S$. The *closure* of A is the set $\bar{A} = \bigcap\{F \subset S \mid A \subset F \text{ and } F \text{ is closed}\}$.

THEOREM 1-3. *Let $\langle S, \tau \rangle$ be a topological space and $A \subset S$. Then $x \in \bar{A}$ iff $x \in G \in \tau$ implies that $G \cap A \neq \varnothing$.*

Proof. Let $x \in \bar{A}$ and $x \in G \in \tau$. Assume that $G \cap A = \varnothing$. This implies that $A \subset S - G$ and $S - G$ is closed. Hence $x \in S - G$. Contradiction. Conversely, suppose that $x \in G \in \tau$ implies $G \cap A \neq \varnothing$. Assume that $x \notin \bar{A}$. Then there is a closed subset F of S such that $F \supset A$ and $x \notin F$. Hence $x \in S - F \in \tau$ and $(S - F) \cap A = \varnothing$. Contradiction.

The points of $\bar{A}$ are called "adherent points" of A. Clearly, $A \subset \bar{A}$ and if A is closed, then $\bar{A} = A$. Otherwise, $S - A$ contains at least one adherent point of A. Such points will be discussed later. Some of the properties of the "closure operator" are contained in a theorem of Kuratowski, which follows.

THEOREM 1-4. *Let $\langle S, \tau \rangle$ be a topological space and $A, B \in 2^S$. Then the following statements are true:*

(K1) $\bar{\varnothing} = \varnothing$.

(K2) $A \subset \bar{A}$.

(K3) $\overline{(\bar{A})} = \bar{A}$.

(K4) $\overline{A \cup B} = \bar{A} \cup \bar{B}$.

Proof

(K1) $\bar{\varnothing} = \bigcap\{F \subset S \mid \varnothing \subset F$ and F is closed$\} = \varnothing$, since $\varnothing$ is closed.

(K2) Since $A \subset \bigcap\{F \subset S \mid A \subset F$ and F is closed$\}$, $A \subset \bar{A}$.

(K3) $\bar{A} = \bigcap\{F \subset S \mid A \subset F$ and F is closed$\}$ is closed by (C2) and $\bar{A} \subset \bar{A}$. Hence $\overline{(\bar{A})} = \bigcap\{F \subset S \mid \bar{A} \subset F$ and F is closed$\} = \bar{A}$.

(K4) Since $A \subset \bar{A}$ and $B \subset \bar{B}$ by (K2), we have $A \cup B \subset \bar{A} \cup \bar{B}$. Since $\bar{A} \cup \bar{B}$ is closed by (C3), we have $\overline{A \cup B} \subset \bar{A} \cup \bar{B}$. Also, $\overline{A \cup B}$ is a closed superset of A and B. Hence $\overline{A \cup B} \supset \bar{A}$ and $\overline{A \cup B} \supset \bar{B}$, which imply that $\overline{A \cup B} \supset \bar{A} \cup \bar{B}$. Thus $\overline{A \cup B} = \bar{A} \cup \bar{B}$.

REMARK. Kuratowski's theorem provides us with a third way to define a topology on a nonempty set S. Let $c : 2^S \to 2^S$ be the closure operator that assigns to each $A \in 2^S$ a subset $c(A) = \bar{A}$ of S such that (K1)–(K4) are satisfied. Define a subset A of S to be closed iff $c(A) = A$. Each such closure operator c determines a unique topology $\tau = \{G \subset S \mid S - G$ is closed$\}$ on S.

EXERCISES

1-1. Let $S \neq \varnothing$ and $\langle \mathscr{L}, \subset \rangle$ be the partially ordered set of all topologies on S under set inclusion. If $\tau_1, \tau_2 \in \mathscr{L}$, define the lattice product $\tau_1 \cdot \tau_2$ to be $\tau_1 \cap \tau_2$ and the lattice sum $\tau_1 + \tau_2$ to be the smallest topology that contains $\tau_1 \cup \tau_2$. Show that under these definitions of lattice sum and product $\langle \mathscr{L}, \subset \rangle$ is a lattice.

1-2. Determine the set $\mathscr{L}$ of all topologies on the set $S = \{a, b, c\}$. Is the resulting lattice $\langle \mathscr{L}, \subset \rangle$ complemented? Distributive?

1-3. Let $\langle S, \tau \rangle$ be a topological space and $A_\alpha \in 2^S$, $\forall \alpha \in \Lambda$. Show that (a) $\overline{\bigcap\{A_\alpha \mid \alpha \in \Lambda\}} \subset \bigcap\{\bar{A}_\alpha \mid \alpha \in \Lambda\}$; (b) $\overline{\bigcup\{A_\alpha \mid \alpha \in \Lambda\}} \supset \bigcup\{\bar{A}_\alpha \mid \alpha \in \Lambda\}$. Give examples to show that the inclusions in (a) and (b) cannot be replaced by equality.

1-4. Let S be an infinite set and $\tau = \{G \subset S \mid S - G \text{ is finite}\} \cup \{\varnothing\}$. Show that τ is a topology on S, which we call the "cofinite topology" on S.

1-5. Define an *interior operator* on S as a function $i : 2^S \to 2^S$ satisfying: (a) $i(S) = S$; (b) $i(A) \subset A$, $\forall A \in 2^S$; (c) $i(i(A)) = i(A)$, $\forall A \in 2^S$; and (d) $i(A \cap B) = i(A) \cap i(B)$. Let $\tau = \{A \subset S \mid i(A) = A\}$. Show that τ is a topology on S.

1-2 Bases and Subbases

In this section we define and investigate two types of structures, bases and subbases, which underlie a topology. Criteria are established for equivalent bases and equivalent subbases. Also, two standard topologies for the set of real numbers are defined.

DEFINITION 1-5. Let $\langle S, \tau \rangle$ be a topological space and $\mathscr{B} \subset 2^S$. $\mathscr{B}$ is a *base for* τ iff τ is the collection consisting of the empty set and all subsets of S that are unions of elements of $\mathscr{B}$.

Our next theorem gives necessary and sufficient conditions under which a subcollection $\mathscr{B}$ of 2^S is a base for a topology τ on S.

THEOREM 1-5. *Let $S \neq \varnothing$ and $\mathscr{B} \subset 2^S$. $\mathscr{B}$ is a base for a topology τ on S iff*

(1) $S = \bigcup\{B \mid B \in \mathscr{B}\}$.

(2) *If $x \in B_1 \cap B_2$, then $\exists B_3 \in \mathscr{B}$ such that $x \in B_3 \subset B_1 \cap B_2$, $\forall B_1, B_2 \in \mathscr{B}$.*

Proof. Suppose that $\mathscr{B}$ is a base for a topology τ on S. Since $S \in \tau$ by (O1), we have $S = \bigcup\{B \mid B \in \mathscr{B}\}$. If $B_1, B_2 \in \mathscr{B}$ and $x \in B_1 \cap B_2$, then $B_1, B_2 \in \tau$ and $B_1 \cap B_2 \in \tau$ by (O3). Thus $B_1 \cap B_2$ is expressible as the union of elements of $\mathscr{B}$, at least one of which must contain x. Hence (1) and (2) are necessary.

Suppose that (1) and (2) hold. Let τ be the collection consisting of $\varnothing$ and all subsets of S that are expressible as unions of elements of $\mathscr{B}$. Then $\varnothing \in \tau$ and $S \in \tau$ by (1). Hence (O1) is satisfied. If $G_\alpha \in \tau$, $\forall \alpha \in \Lambda$, and $x \in \bigcup\{G_\alpha \mid \alpha \in \Lambda\}$, then $x \in$ some G_α. Since each G_α is a union of elements of $\mathscr{B}$, $\exists B \in \mathscr{B}$ such that $x \in B \subset G_\alpha \subset \bigcup\{G_\alpha \mid \alpha \in \Lambda\}$. Thus $\bigcup\{G_\alpha \mid \alpha \in \Lambda\}$ is expressible as the union of elements of $\mathscr{B}$ and hence belongs to τ. Axiom (O2) is thus satisfied. To show that τ is closed under the formation of finite intersections [axiom (O3)], we need only show that the intersection of any two elements of τ is expressible as the union of elements of $\mathscr{B}$. Let $G_1, G_2 \in \tau$ and $x \in G_1 \cap G_2$. Then $\exists B_1, B_2 \in \mathscr{B}$ with $x \in B_1 \subset G_1$ and $x \in B_2 \subset G_2$. This implies that $x \in B_1 \cap B_2 \subset G_1 \cap G_2$. Condition (2) implies that $\exists B_3 \in \mathscr{B}$ such that $x \in B_3 \subset B_1 \cap B_2 \subset G_1 \cap G_2$. Thus $G_1 \cap G_2$ is expressible as the union of elements of $\mathscr{B}$, and hence is a member of τ.

EXAMPLE 1-2. Let R be the set of real numbers and $\mathscr{B}_1 = \{(a, b) \mid a, b \in R$ with $a < b\}$, where $(a, b) = \{x \in R \mid a < x < b\}$. $\mathscr{B}_1$ is a base for a topology ξ on R, which we call the "interval (Euclidean) topology." Let $\mathscr{B}_2 = \{[a, b) \mid a, b \in R$ with $a < b\}$, where $[a, b) = \{x \in R \mid a \leq x < b\}$. $\mathscr{B}_2$ is a base for a topology $\mathscr{L}$ on R, which we call the "lower limit topology." Similarly, if we let $\mathscr{B}_3 = \{(a, b] \mid a, b \in R$ with $a < b\}$, where $(a, b] = \{x \in R \mid a < x \leq b\}$, then $\mathscr{B}_3$ is a base for a topology $\mathscr{U}$ on R, which we call the "upper limit topology." Since $\langle R, \mathscr{L}\rangle$ and $\langle R, \mathscr{U}\rangle$ have the same topological properties, we shall only investigate those of $\langle R, \mathscr{L}\rangle$. However, $\langle R, \xi\rangle$ and $\langle R, \mathscr{L}\rangle$ have quite different topological properties, as we shall see later.

DEFINITION 1-6. Let $\mathscr{B}_1 \subset 2^S$ and $\mathscr{B}_2 \subset 2^S$. Then $\mathscr{B}_1$ and $\mathscr{B}_2$ are *equivalent bases* iff they are bases for the same topology on S. Criteria for two subcollections of 2^S to be equivalent bases are established in the next theorem.

THEOREM 1-6. *Let $\langle S, \tau\rangle$ be a topological space, $\mathscr{B}_1$ a base for τ, and $\mathscr{B}_2 \subset 2^S$. If*

(1) *$x \in B_1 \in \mathscr{B}_1$ implies that $\exists B_2 \in \mathscr{B}_2$ such that $x \in B_2 \subset B_1$, and*

(2) *$x \in B_2 \in \mathscr{B}_2$ implies that $\exists B_1 \in \mathscr{B}_1$ such that $x \in B_1 \subset B_2$,*

then $\mathscr{B}_2$ is also a base for τ, and $\mathscr{B}_1$ and $\mathscr{B}_2$ are thus equivalent.

Proof. Let $\varnothing \neq G \in \tau$. Thus G is the union of elements of $\mathscr{B}_1$ by Definition 1-5. Condition (1) implies that each element of $\mathscr{B}_1$ is expressible as the union of elements of $\mathscr{B}_2$. Hence G is the union of elements of $\mathscr{B}_2$. Now let G be any union of elements of $\mathscr{B}_2$. Condition (2) implies that each element of $\mathscr{B}_2$ is expressible as the union of elements of $\mathscr{B}_1$. Hence G is the union of elements of $\mathscr{B}_1$ and $G \in \tau$. Thus $\mathscr{B}_2$ is a base for τ.

EXAMPLE 1-3. Let $S = \{\langle x, y\rangle \mid x, y$ are real numbers$\}$. The distance between $\langle x_1, y_1\rangle$ and $\langle x_2, y_2\rangle$ in S is given by $\sqrt{(x_2 - x_1)^2 + (y_2 - y_1)^2}$. For each $\langle x_0, y_0\rangle \in S$ and every $\varepsilon > 0$, let $S(\langle x_0, y_0\rangle; \varepsilon) = \{\langle x, y\rangle \in S \mid \sqrt{(x - x_0)^2 + (y - y_0)^2} < \varepsilon\}$. The collection $\mathscr{B}_1 = \{S(\langle x_0, y_0\rangle; \varepsilon) \mid \langle x_0, y_0\rangle \in S, \varepsilon > 0\}$ is a base for a topology ξ^2 on S, which is called the Euclidean topology. Geometrically, $\mathscr{B}_1$ is the collection of all open circular disks in the plane. An equivalent base is $\mathscr{B}_2 = \{\{\langle x, y\rangle \in S \mid a < x < b, c < y < d\} \mid a, b, c, d \in R\}$, the collection of all open rectangular regions in the plane.

DEFINITION 1-7. Let $\langle S, \tau\rangle$ be a topological space and $\mathscr{S} \subset 2^S$. If $\mathscr{B}$ is the collection of all intersections of finitely many elements of $\mathscr{S}$, then $\mathscr{S}$ is a *subbase* for τ iff $\mathscr{B}$ is a base for τ.

THEOREM 1-7. *If* $\mathscr{S} \subset 2^S$ *and* $\bigcup\{G \mid G \in \mathscr{S}\} = S$, *then* $\mathscr{S}$ *is a subbase for a unique topology* τ *on* S.

Proof. Let $\mathscr{B}$ be the collection of all intersections of finitely many elements of $\mathscr{S}$. Thus $\mathscr{S} \subset \mathscr{B}$. Let $x \in S$ be arbitrary. Since $\bigcup\{G \mid G \in S\} = S$, $\exists G \in \mathscr{S}$ such that $x \in G$. Hence condition (1) of Theorem 1-5 is satisfied. Now let B_1 and B_2 be elements of $\mathscr{B}$ and $x \in B_1 \cap B_2$. Since B_1 and B_2 are intersections of finitely many elements of $\mathscr{S}$, $B_1 \cap B_2$ is the intersection of finitely many elements of $\mathscr{S}$ and thus belongs to $\mathscr{B}$. Hence condition (2) of Theorem 1-5 is satisfied also, and $\mathscr{B}$ is a base for a unique topology τ on S. Using Definition 1-7, we conclude that $\mathscr{S}$ is a subbase for τ.

EXAMPLE 1-4. Let $\mathscr{S} = \{(-\infty, b) \mid b \in R\} \cup \{(a, +\infty) \mid a \in R\}$. Since $\mathscr{B} = \{(a, b) \mid a, b \in R, a < b\} \cup \mathscr{S}$ is a base for the interval topology ξ on R, $\mathscr{S}$ is a subbase for ξ by Definition 1-7.

EXERCISES

1-6. Verify that the collections $\mathscr{B}_1$ and $\mathscr{B}_2$ described in Example 1-2 are bases for topologies on the set R of real numbers.

1-7. Verify that the collections $\mathscr{B}_1$ and $\mathscr{B}_2$ described in Example 1-3 are equivalent bases.

1-8. Let $\mathscr{S}_1 \subset 2^S$ and $\mathscr{S}_2 \subset 2^S$ with $\bigcup\{G_1 \mid G_1 \in \mathscr{S}_1\} = S = \bigcup\{G_2 \mid G_2 \in \mathscr{S}_2\}$. If (1) $x \in G_1 \in \mathscr{S}_1$ implies that $\exists G_2 \in \mathscr{S}_2$ with $x \in G_2 \subset G_1$, and (2) $x \in G_2 \in \mathscr{S}_2$ implies that $\exists G_1 \in \mathscr{S}_1$ with $x \in G_1 \subset G_2$, then $\mathscr{S}_1$ and $\mathscr{S}_2$ are equivalent subbases (i.e., subbases for the same topology on S).

1-3 Limit Points, Boundary Points, and Sequential Limits

In Section 1-1 we referred to $\bar{A}$ both as the closure of A and as the set of adherent points of A. In this section we investigate three special categories of adherent points—limit points, boundary points, and sequential limits.

DEFINITION 1-8. Let $\langle S, \tau \rangle$ be a topological space. Then $x \in S$ is a *limit point* of $A \subset S$ iff $x \in G \in \tau$ implies that $(G - \{x\}) \cap A \neq \emptyset$. The set of limit points of A is called the *derived set* of A and denoted by A'.

DEFINITION 1-9. Let $\langle S, \tau \rangle$ be a topological space. Then $x \in S$ is a *boundary point* of $A \subset S$ iff $x \in G \in \tau$ implies that $G \cap A \neq \emptyset$ and $G \cap (S - A) \neq \emptyset$. The set of boundary points of A is called the *boundary* of A and denoted by $B(A)$.

THEOREM 1-8. *If* $\langle S, \tau \rangle$ *is a topological space and* $A \subset S$, *then* $\bar{A} = A \cup A' = A \cup B(A)$.

Proof. If $x \notin \bar{A}$, then $\exists G \in \tau$ such that $x \in G$ and $G \cap A = \emptyset$. Hence $x \notin A$ and $x \notin A'$ (i.e., $x \notin A \cup A'$). Thus $A \cup A' \subset \bar{A}$. On the other hand, $x \notin A \cup A'$ implies that $\exists G \in \tau$ such that $x \in G$ and $G \cap A = \emptyset$. Hence $x \notin \bar{A}$. Thus $\bar{A} \subset A \cup A'$. In view of Definition 0-2, we have $\bar{A} = A \cup A'$. The reader may construct a similar argument to show that $\bar{A} = A \cup B(A)$.

EXAMPLE 1-5. This example shows that, generally speaking, $A' \not\subset B(A)$ and $B(A) \not\subset A'$ even though $A \cup A' = A \cup B(A)$ from Theorem 1-8. Let $\langle R, \xi \rangle$ be the real line with the interval topology and $A = (0, 1) \cup \{2\}$. Then $\bar{A} = [0, 1] \cup \{2\}$, $A' = [0, 1]$, and $B(A) = \{0, 1, 2\}$. $B(A) \not\subset A'$, since $2 \notin A'$, and $A' \not\subset B(A)$, since no point of the open interval $(0, 1)$ is contained in $B(A)$.

DEFINITION 1-10. If $\langle S, \tau \rangle$ is a topological space and $A, B \in 2^S$, then A is *dense in* B iff $B \subset \bar{A}$. This implies that $A \subset S$ is *dense in* S iff $G \cap A \neq \emptyset, \forall G \in \tau - \{\emptyset\}$.

DEFINITION 1-11. If $\langle S, \tau \rangle$ is a topological space and $A \subset S$, then A is *nowhere dense in* S iff $\bar{A}$ contains no member of $\tau - \{\emptyset\}$.

EXAMPLE 1-6. Let $\langle R, \xi \rangle$ be the real line with the interval topology. The set of rationals is dense in R, since any open interval containing an irrational contains infinitely many rationals. The set of integers is nowhere dense in R, since it contains no intervals and hence no elements of $\xi - \{\varnothing\}$. Notice that each rational (and each real) is a limit point of the set R of reals. Indeed, since R is closed, we have $\bar{R} = R = R'$.

DEFINITION 1-12. If $\langle S, \tau \rangle$ is a topological space and $A \subset S$, then A is *perfect* iff A is closed and $A \subset A'$ (i.e., $\bar{A} = A = A'$).

DEFINITION 1-13. Let $\langle S, \tau \rangle$ be a topological space, $\{x_n\}_{n \in I^+}$ a sequence in S, and $x \in S$. We shall say that $\{x_n\}_{n \in I^+}$ *converges to* x and write $x_n \to x$ iff $x \in G \in \tau$ implies that $\exists N \in I^+$ such that $x_n \in G$, $\forall n \geq N$. As a sequential limit, x is also designated by "$\lim_{n \to \infty} x_n$."

A sequence may have no limit, a unique limit, or several limits, depending upon the topology. We describe now a sequence that has infinitely many limits.

EXAMPLE 1-7. Let R be the set of real numbers and let τ be the cofinite topology on R (see Exercise 1-4). Let $x_n = n$, $\forall n \in I^+$. If $x \in R$ and $x \in R \in \tau$, then $R - G$ is finite and $x_n \in G$, $\forall n \geq$ some $N \in I^+$. Thus $x_n \to x$, $\forall x \in R$.

The theorem that follows indicates the way in which sequential limits are related to adherent points and limit points.

THEOREM 1-9. *Let $\langle S, \tau \rangle$ be a topological space, $A \subset S$ and $x \in S$.*
(1) *If $\{x_n\}_{n \in I^+}$ is a sequence in A such that $x_n \to x$, then $x \in \bar{A}$.*
(2) *If $\{x_n\}_{n \in I^+}$ is a sequence of distinct points in A such that $x_n \to x$, then $x \in A'$.*

Proof

(1) Let $\{x_n\}_{n \in I^+}$ be a sequence in A converging to x and $x \in G \in \tau$. By Definition 1-13, $\exists N \in I^+$ such that $x_n \in G$, $\forall n \geq N$. Since $x_n \in A$, $\forall n \in I^+$, we have $G \cap A \neq \varnothing$ and $x \in \bar{A}$ by Theorem 1-3.

The proof of (2) is left as an exercise for the reader.

Generally speaking, sequential limits are not necessarily limit points, and limit points are not necessarily sequential limits. The example that follows illustrates this fact.

EXAMPLE 1-8. Let $S = \{a, b, c\}$ and $\tau = \{\varnothing, \{a, b\}, \{c\}, S\}$. Let $x_1 = a$, $x_2 = b$, and $x_n = c$, $\forall n \geq 3$. Clearly, $x_n \to c$, but $c \notin \{a, b, c\}'$, since $c \in \{c\} \in \tau$ and $(\{c\} - \{c\}) \cap \{a, b, c\} = \varnothing$. Moreover, $a, b \in \{a, b, c\}'$ by Definition 1-8, but $x_n \nrightarrow a$ and $x_n \nrightarrow b$, since $a, b \in \{a, b\} \in \tau$ and $x_n \notin \{a, b\}$, $\forall n \geq 3$.

EXERCISES

1-9. Let $\langle S, \tau \rangle$ be a topological space and $A \subset S$. Show that A is closed iff $A' \subset A$.

1-10. Let $\langle S, \tau \rangle$ be a topological space and $A \subset B \subset S$. Show that $A' \subset B'$.

1-11. Prove Theorem 1-9(2).

1-12. Let R be the set of real numbers and $\tau = \{\varnothing\} \cup \{(a, +\infty) \mid a \in R\} \cup \{R\}$. Verify that τ is a topology on R and establish in $\langle R, \tau \rangle$ the following sequential convergence and divergence:

(a) If $x_n = n$, $\forall n \in I^+$, then $x_n \to x$, $\forall x \in R$.
(b) If $x_n = -n$, $\forall n \in I^+$, then $\{x_n\}_{n \in I^+}$ does not converge in R.
(c) If $x_n = (-1)^n$, $\forall n \in I^+$, then $x_n \to x$, $\forall x \leq -1$.

1-13. If $p \geq 2$ is an integer, a "p-adic rational" is a real number $r = k \cdot p^{-n}$ for some nonnegative integer k and positive integer n. Show that the set of p-adic rationals in $I^1 = [0, 1]$ is dense in I^1.

1-14. The *Cantor ternary set* K is the set of all $x \in [0, 1]$ having a ternary expansion $x = t_1 \cdot 3^{-1} + t_2 \cdot 3^{-2} + \cdots + t_n \cdot 3^{-n} + \cdots$ with $t_n \neq 1$, $\forall n \in I^+$. Intuitively, K can be thought of as the set obtained from $[0, 1]$ following the successive removal of all open middle thirds. Figure 1-1 illustrates the first two stages in this geometrical construction of K. Show that K is uncountable, perfect, and nowhere dense in $[0, 1]$.

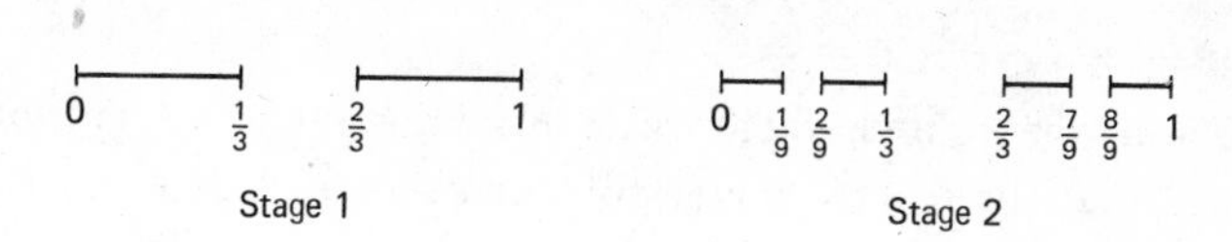

FIGURE 1-1

1-4 Continuity, Homeomorphisms, and Topological Properties

In this section we discuss the continuity of a function from one topological space into another. We define a homeomorphism (or topological mapping) as a 1-1 continuous function whose inverse is also continuous. Properties of topological spaces that are invariant (preserved) under homeomorphisms are known as topological properties and are the objects of our interest throughout the remainder of this book.

DEFINITION 1-14. $f: \langle S, \tau_1 \rangle \to \langle T, \tau_2 \rangle$ is *continuous at* $x \in S$ iff $f(x) \in U \in \tau_2$ implies that $\exists G \in \tau_1$ such that $x \in G$ and $f(G) \subset U$. f is *continuous* iff f is continuous at each $x \in S$. A continuous function is also known as a *mapping*.

THEOREM 1-10. *If* $f: \langle S, \tau_1 \rangle \xrightarrow{onto} \langle T, \tau_2 \rangle$ *is a function, then the following statements are equivalent:*

(1) $f^{-1}(C)$ *is closed in S whenever C is closed in T.*

(2) $f^{-1}(U) \in \tau_1$, $\forall U \in \tau_2$.

(3) f *is continuous.*

(4) $f(\bar{A}) \subset \overline{f(A)}$, $\forall A \subset S$.

Proof. We demonstrate the equivalence by establishing the cycle of implications: (1) $\to$ (2) $\to$ (3) $\to$ (4) $\to$ (1).

(1) $\to$ (2). Let $U \in \tau_2$. Then $T - U$ is closed, which implies that $f^{-1}(T - U)$ is closed. Since $f^{-1}(T - U) = f^{-1}(T) - f^{-1}(U) = S - f^{-1}(U)$, we have that $f^{-1}(U) \in \tau_1$.

(2) $\to$ (3). Let $x \in S$ and $f(x) \in U \in \tau_2$. Then $x \in f^{-1}(U) \in \tau_1$, and since $f(f^{-1}(U)) \subset U$, f is continuous at x and thus continuous, since x was arbitrary.

(3) $\to$ (4). Let $A \subset S$. If $y \in f(\bar{A})$ and $y \in U \in \tau_2$, then $y = f(x)$ for some $x \in \bar{A}$. Since f is continuous, $\exists V \in \tau_1$ such that $x \in V$ and $f(V) \subset U$. Also, $x \in \bar{A}$ implies that $\exists p \in V \cap A$. Thus $f(p) \in f(V) \cap f(A) \subset U \cap f(A)$. Hence $y \in \overline{f(A)}$ and $f(\bar{A}) \subset \overline{f(A)}$.

(4) $\to$ (1). Let C be a closed subset of T. Then $f(\overline{f^{-1}(C)}) \subset \overline{f(f^{-1}(C))} \subset \bar{C} = C$. This implies that $\overline{f^{-1}(C)} \subset f^{-1}(C)$, and so $f^{-1}(C)$ is closed.

The previous theorem characterizes continuous functions as those having the property that the inverse images of open sets are open and the inverse images of closed sets are closed. However, continuous functions do *not* necessarily map open sets onto open sets and closed sets onto closed sets, as the following example illustrates.

EXAMPLE 1-9. Let $S = T = \{a, b, c\}$, $\tau_1 = \{\varnothing, \{a\}, S\}$, and $\tau_2 = \{\varnothing, \{a\}, \{b, c\}, T\}$. Let $f(x) = b$, $\forall x \in S$. f is continuous and $f^{-1}(\varnothing) = f^{-1}(\{a\}) = \varnothing \in \tau_1$ and $f^{-1}(\{b, c\}) = f^{-1}(T) = S \in \tau_1$. However, $\{a\} \in \tau_1$ and $f(\{a\}) = \{b\} \notin \tau_2$. Also, $\{b, c\}$ is closed in S and $f(\{b, c\}) = \{b\}$ is not closed in T.

DEFINITION 1-15. Let $f: \langle S, \tau_1 \rangle \to \langle T, \tau_2 \rangle$ be a function:
(1) f is *open* iff $G \in \tau_1$ implies that $f(G) \in \tau_2$.
(2) f is *closed* iff C closed in S implies that $f(C)$ is closed in T.

Our next theorem states that continuous functions preserve sequential convergence. Although the converse is false, as the example that follows shows, we shall later obtain a converse of this theorem in the context of "first countable" spaces.

THEOREM 1-11. *If* $f: \langle S, \tau_1 \rangle \to \langle T, \tau_2 \rangle$ *is continuous and* $x_n \to x$ *in* S, *then* $f(x_n) \to f(x)$ *in* T.

Proof. Let $f(x) \in U \in \tau_2$. Then $x \in f^{-1}(U) \in \tau_1$ by Theorem 1-10. Hence $\exists N \in I^+$ such that $x_n \in f^{-1}(U)$, $\forall n \geq N$, by Definition 1-13. Thus $f(x_n) \in U$, $\forall n \geq N$, and $f(x_n) \to f(x)$.

EXAMPLE 1-10. Let S be the set of real numbers and $\tau_1 = \{\varnothing\} \cup \{G \subset S \mid S - G \text{ is countable}\}$. τ_1 is a topology on S called "the co-countable topology." Let $T = [0, 1]$ and $\tau_2 = \{G \cap [0, 1] \mid G \in \tau_1\}$. Then τ_2 is a topology on T known as the "subspace topology" induced on T by τ_1. Let

$$f(x) = \begin{cases} x & \text{if } x \in [0, 1] \\ 0 & \text{otherwise.} \end{cases}$$

Then f is *not* continuous, since $(0, 1) \in \tau_2$, but $f^{-1}[(0, 1)] = (0, 1) \notin \tau_1$ as $R - (0, 1)$ is uncountable. However, $x_n \to x$ in S iff $x_n = x$, $\forall n \geq$ some $N \in I^+$, iff $f(x_n) = f(x)$, $\forall n \geq N$, iff $f(x_n) \to f(x)$.

DEFINITION 1-16. A function $h: \langle S, \tau_1 \rangle \xrightarrow{\text{onto}} \langle T, \tau_2 \rangle$ is a *homeomorphism* (*topological mapping*) iff h is 1-1, and h and h^{-1} are continuous.

DEFINITION 1-17. A property P of a topological space is a *topological property* iff P is invariant (preserved) under homeomorphisms.

THEOREM 1-12. *If* $h: \langle S, \tau_1 \rangle \xrightarrow{\text{onto}} \langle T, \tau_2 \rangle$ *is* 1-1, *the following statements are equivalent:*
(1) *h is a homeomorphism.*
(2) *h is open and continuous.*
(3) *h is closed and continuous.*
(4) $h(\bar{A}) = \overline{h(A)}$, $\forall A \subset S$.

Proof. Left to the reader as an exercise.

REMARK. The relation "$\langle S, \tau_1 \rangle$ is homeomorphic to $\langle T, \tau_2 \rangle$" is an equivalence relation on the collection of all topological spaces. In particular, the doughnut and coffee cup belong to the same equivalence class.

EXERCISES

1-15. Let $f: \langle S, \tau_1 \rangle \to \langle T, \tau_2 \rangle$ and $\mathscr{B}_1$ and $\mathscr{B}_2$ be bases for τ_1 and τ_2, respectively. Show that f is continuous at $x \in S$ iff $f(x) \in B_2 \in \mathscr{B}_2$ implies that $\exists B_1 \in \mathscr{B}_1$ such that $x \in B_1$ and $f(B_1) \subset B_2$.

1-16. If $f_1 : \langle S_1, \tau_1 \rangle \to \langle S_2, \tau_2 \rangle$ and $f_2 : \langle S_2, \tau_2 \rangle \to \langle S_3, \tau_3 \rangle$ are continuous, show that $f_2 \circ f_1 : \langle S_1, \tau_1 \rangle \to \langle S_3, \tau_3 \rangle$ is also continuous.

1-17. Prove Theorem 1-12.

1-18. Show that the collection of all homeomorphisms of a topological space $\langle S, \tau \rangle$ onto itself is a group under the "composition of homeomorphisms."

1-19. For each $p \in R$, let $f(p) = L_p = \{\langle x, y \rangle \in R \times R \mid y = px\}$, and let $S = \{L_p \mid p \in R\}$.

(a) Show that $f: R \to S$ is 1-1 and onto.

(b) Let $(L_a, L_b) = \{L_p \in S \mid a < p < b\}$. Sketch (L_{-1}, L_1) and $f^{-1}[(L_{-1}, L_1)]$.

(c) Let $\tau = \{G \subset S \mid f^{-1}(G) \in \xi$ (interval topology on R)$\}$. Show that τ is a topology on S.

(d) Show that $\langle R, \xi \rangle$ and $\langle S, \tau \rangle$ are homeomorphic.

1-5 Subspaces and Product Spaces

In this section we introduce the notions of "subspace" and "product space." Related to these notions are the concepts of "hereditary property" and "productive property." Also, the "projection mappings" of a product space onto its factor spaces are defined and shown to be continuous and open.

DEFINITION 1-18. Let $\langle S, \tau \rangle$ be a topological space and $\varnothing \neq A \subset S$. The *subspace* (*relative*) *topology* on A is $\tau/A = \{A \cap G \mid G \in \tau\}$. $\langle A, \tau/A \rangle$ is a *subspace* of $\langle S, \tau \rangle$.

The reader should verify that τ/A is a topology on A by showing that axioms (O1)–(O3) of Definition 1-1 are satisfied. We also leave as an exercise the proof of the theorem that follows. It states that $\mathscr{B}/A$ is a base for the subspace topology τ/A on $A \subset S$ whenever $\mathscr{B}$ is a base for τ.

THEOREM 1-13. *Let $\langle S, \tau\rangle$ be a topological space and $\mathscr{B}$ a base for τ. If $\varnothing \neq A \subset S$, then $\mathscr{B}/A = \{A \cap B \mid B \in \mathscr{B}\}$ is a base for τ/A.*

EXAMPLE 1-11. Consider the real line $E^1 = \langle R, \xi\rangle$. Let $A = [0, 1]$ and $I^1 = \langle A, \xi/A\rangle$. A base for ξ/A is the collection $\{[0, c) \mid 0 < c \leq 1\} \cup \{(d, 1] \mid 0 \leq d < 1\}$. Any homeomorphic image of I^1 is called an *arc* or (closed) 1-*cell.*

DEFINITION 1-19. A topological property P is *hereditary* iff each subspace of a space having property P also has property P.

Next we define the "product space" for a collection $\{\langle S_\alpha, \tau_\alpha\rangle \mid \alpha \in \Lambda\}$ of topological spaces. We begin with definitions of the general Cartesian product $\prod\{S_\alpha \mid \alpha \in \Lambda\}$ and the "projection mappings" $\pi_\alpha : \prod\{S_\alpha \mid \alpha \in \Lambda\} \to S_\alpha$. Then we assign to $\prod\{S_\alpha \mid \alpha \in \Lambda\}$ the "Tychonoff product topology."

DEFINITION 1-20. Let $S_\alpha \neq \varnothing$, $\forall \alpha \in \Lambda$.

(1) The Cartesian product $\prod\{S_\alpha \mid \alpha \in \Lambda\} = \{$all functions $x : \Lambda \to \bigcup\{S_\alpha \mid \alpha \in \Lambda\} \mid x(\alpha) \in S_\alpha, \forall \alpha \in \Lambda\}$.

(2) For each $\alpha \in \Lambda$, the *projection mapping* $\pi_\alpha : \prod\{S_\alpha \mid \alpha \in \Lambda\} \to S_\alpha$ is given by $\pi_\alpha(x) = x(\alpha)$, $\forall x \in \prod\{S_\alpha \mid \alpha \in \Lambda\}$.

DEFINITION 1-21. Let $\langle S_\alpha, \tau_\alpha\rangle$ be a topological space $\forall \alpha \in \Lambda$, and let $S = \prod\{S_\alpha \mid \alpha \in \Lambda\}$. Let $\mathscr{S} = \{\pi_\alpha^{-1}(G_\alpha) \mid G_\alpha \in \tau_\alpha, \forall \alpha \in \Lambda\}$; $\mathscr{S}$ is a subbase for the *Tychonoff product topology* τ on S. Also, $\langle S, \tau\rangle$ is the *product space* $\prod\{\langle S_\alpha, \tau_\alpha\rangle \mid \alpha \in \Lambda\}$ and $\langle S_\alpha, \tau_\alpha\rangle$ is its *αth factor (coordinate) space.*

EXAMPLE 1-12. Let $S_1 = S_2 = \{a, b, c\}$, $\tau_1 = \{\varnothing, \{a\}, \{b, c\}, S_1\}$, and $\tau_2 = \{\varnothing, \{a, b\}, \{c\}, S_2\}$. Then, $S_1 \times S_2 = \{\langle a, a\rangle, \langle a, b\rangle, \langle a, c\rangle, \langle b, a\rangle, \langle b, b\rangle, \langle b, c\rangle, \langle c, a\rangle, \langle c, b\rangle, \langle c, c\rangle\}$. The product topology $\tau_1 \times \tau_2$ on $S_1 \times S_2$ has by Definition 1-21 a subbase $\mathscr{S} = \{\pi_1^{-1}(\{a\}), \pi_1^{-1}(\{b, c\}), \pi_2^{-1}(\{a, b\}), \pi_2^{-1}(\{c\}), S_1 \times S_2\} = \{\{\langle a, a\rangle, \langle a, b\rangle, \langle a, c\rangle\}, \{\langle b, a\rangle, \langle b, b\rangle, \langle b, c\rangle, \langle c, a\rangle, \langle c, b\rangle, \langle c, c\rangle\}, \{\langle a, a\rangle, \langle b, a\rangle, \langle c, a\rangle, \langle a, b\rangle, \langle b, b\rangle, \langle c, b\rangle\}, \{\langle a, c\rangle, \langle b, c\rangle, \langle c, c\rangle\}, S_1 \times S_2\}$. A base for $\tau_1 \times \tau_2$ is $\mathscr{B} = \mathscr{S} \cup \{\{\langle a, a\rangle, \langle a, b\rangle\}, \{\langle a, c\rangle\}, \{\langle b, a\rangle, \langle b, b\rangle, \langle c, a\rangle, \langle c, b\rangle\}, \{\langle b, c\rangle, \langle c, c\rangle\}\}$. We leave as an exercise for the reader the determination of the elements of $\tau_1 \times \tau_2$, where $\tau_1 \times \tau_2$ is *not* to be construed as the Cartesian product.

THEOREM 1-14. *The projections of a product space onto its factor spaces are continuous and open.*

Proof. Let $\langle S, \tau\rangle = \prod\{S_\alpha, \tau_\alpha\rangle \mid \alpha \in \Lambda\}$ and $\pi_\alpha : S \xrightarrow{\text{onto}} S_\alpha$. Since $\mathscr{S} = \{\pi_\alpha^{-1}(G_\alpha) \mid G_\alpha \in \tau_\alpha, \alpha \in \Lambda\}$ is a subbase for τ, $G_\alpha \in \tau_\alpha$ implies that $\pi_\alpha^{-1}(G_\alpha) \in$

$\mathscr{S} \subset \tau$, and π_α is continuous. Let $\mathscr{B}$ be the base for τ consisting of all intersections of finitely many elements of $\mathscr{S}$. To show that π_α is open, it suffices to show that $\pi_\alpha(B) \in \tau_\alpha$, $\forall B \in \mathscr{B}$. Let $B = \bigcap\{\pi_{\alpha_i}^{-1}(G_{\alpha_i}) \mid G_{\alpha_i} \in \tau_{\alpha_i}, i = 1, 2, \ldots, n\} \in \mathscr{B}$. If $\alpha \in \{\alpha_1, \alpha_2, \ldots, \alpha_n\}$, then $\pi_\alpha(B) = G_\alpha$. If $\alpha \notin \{\alpha_1, \alpha_2, \ldots, \alpha_n\}$, then $\pi_\alpha(B) = S_\alpha$. In either case, $\pi_\alpha(B) \in \tau_\alpha$ and π_α is open.

EXAMPLE 1-13. If $\langle S_i, \tau_i\rangle = E^1 = \langle R, \xi\rangle$ for $i = 1, 2, \ldots, n$, then the product space $\langle S, \tau\rangle = \prod\{\langle S_i, \tau_i\rangle \mid i = 1, 2, \ldots, n\}$ is denoted by E^n and called "Euclidean n-space." It has a very important subspace $I^n = \langle A, \tau/A\rangle = \prod\{\langle A_i, \tau_i/A_i\rangle \mid i = 1, 2, 3, \ldots, n\}$, where $\langle A_i, \tau_i/A_i\rangle = I^1 = \langle [0, 1], \xi/[0, 1]\rangle$ for $i = 1, 2, \ldots, n$. I^n is called the *n-dimensional unit cube*, and any homeomorphic image of it is called an (closed) *n-cell*.

EXAMPLE 1-14. In Example 1-3 we described the Euclidean topology on $R \times R$. If we let $\mathscr{B} = \{[a, b) \times [c, d) \mid a < b, c < d$ and $a, b, c, d \in R\}$, then $\mathscr{B}$ is a base for a topology $\mathscr{L} \times \mathscr{L}$ on R which we call the *left-lower box topology*. We shall see later that $\mathscr{L} \times \mathscr{L}$ is radically different from $\xi \times \xi$.

DEFINITION 1-22. A topological property P is *productive* iff whenever $\langle S_\alpha, \tau_\alpha\rangle$ has property P, $\forall \alpha \in \Lambda$, $\prod\{\langle S_\alpha, \tau_\alpha\rangle \mid \alpha \in \Lambda\}$ also has property P.

EXERCISES

1-20. If $\langle S, \tau\rangle$ is a topological space and $\varnothing \neq A \subset S$, verify that $\tau/A = \{A \cap G \mid G \in \tau\}$ is a topology on A.

1-21. Prove Theorem 1-13.

1-22. Let $\langle S, \tau\rangle$ be a topological space and $\varnothing \neq F \subset S$. We say that $A \subset F$ is *closed in F* iff $F - A \in \tau/F$. Show that A is closed in F iff there exists a closed subset C of S such that $A = C \cap F$. Show also that if F is closed, then A is closed in F iff A is closed.

1-23. Let $f: \langle S, \tau_1\rangle \to \langle T, \tau_2\rangle$ be continuous. Show that $f/A : \langle A, \tau_1/A\rangle \to \langle T, \tau_2\rangle$ is continuous $\forall A \subset S$.

1-24. Let $\langle S, \tau\rangle$ be a topological space and $\varnothing \neq C \subset A \cup B \subset S$. Show that $C \in \tau/A \cup B$ iff $C \cap A \in \tau/A$ and $C \cap B \in \tau/B$.

1-25. Show that a function $f: \langle T, \tau\rangle \to \prod\{\langle S_\alpha, \tau_\alpha\rangle \mid \alpha \in \Lambda\}$ is continuous iff $\pi_\alpha \circ f: \langle T, \tau\rangle \to \langle S_\alpha, \tau_\alpha\rangle$ is continuous $\forall \alpha \in \Lambda$.

1-26. Give $S = \{a, b, c\}$ a topology $\tau_1 = \{\varnothing, \{a\}, \{b, c\}, S\}$ and give $T = \{1, 2\}$ a topology $\tau_2 = \{\varnothing, \{1\}, T\}$. Determine a subbase $\mathscr{S}$ and a base $\mathscr{B}$ for the product topology $\tau_1 \times \tau_2$ on $S \times T$.

1-27. Let $\langle S, \tau\rangle = \prod\{S_\alpha, \tau_\alpha\rangle \mid \alpha \in \Lambda\}$ and let $\mathscr{B}_\alpha$ be a base for τ_α, $\forall \alpha \in \Lambda$. Show that $\mathscr{S}^* = \{\pi_\alpha^{-1}(B_\alpha) \mid B_\alpha \in \mathscr{B}_\alpha, \alpha \in \Lambda\}$ is a subbase for τ.

1-28. If $\langle S_i, \tau_i\rangle$ is a topological space for $i = 1, 2, \ldots, n$, then $\mathscr{B}^* = \{G_1 \times \cdots \times G_n \mid G_i \in \tau_i, i = 1, 2, \ldots, n\}$ is a base for the "box topology" on $S_1 \times \cdots \times S_n$. Prove the Tychonoff product topology and the "box topology" coincide in the case of finite products.

1-29. If $\mathscr{B}_1$ is a base for a topology τ_1 on S_1 and $\mathscr{B}_2$ is a base for a topology τ_2 on S_2, show that $\mathscr{B}_1 \times \mathscr{B}_2 = \{B_1 \times B_2 \mid B_1 \in \mathscr{B}_1 \text{ and } B_2 \in \mathscr{B}_2\}$ is a base for the product (box) topology on $S_1 \times S_2$.

1-6 Separable Spaces

Spaces that contain countable dense subspaces are called "separable"; for example, $\langle R, \xi\rangle$ is separable, since the rationals are countable and dense in the reals with respect to the interval topology. Because homeomorphisms are 1-1 and preserve adherent points, it follows that the property of being "separable" is a topological property.

DEFINITION 1-23. $\langle S, \tau\rangle$ is *separable* iff S contains a countable subset D that is dense in S (i.e., $\bar{D} = S$).

The set of reals under the discrete topology is *not* separable, since no proper subspace is dense. Thus separability depends primarily on the topology τ rather than on the set S. We show by example that separability is not hereditary. Although separability is not productive either, we do prove that the product of a countable number of separable spaces is separable.

EXAMPLE 1-15. Consider $R \times R$ with the "left-lower box topology" described in Example 1-14. Since $\langle R, \mathscr{L}\rangle$ is separable (let D = {rationals}), $\langle R \times R, \mathscr{L} \times \mathscr{L}\rangle$ is also separable by Theorem 1-16, which follows. Consider the subset $L = \{\langle x, -x\rangle \mid x \in R\}$ with the subspace topology. Since this subspace topology on L is discrete, the subspace L is not separable.

THEOREM 1-15. *Every open subspace of a separable space is separable.*
Proof. Left as an exercise for the reader.

THEOREM 1-16. *If $\langle S_n, \tau_n\rangle$ is separable $\forall n \in I^+$, then $\prod\{\langle S_n, \tau_n\rangle \mid n \in I^+\}$ is also separable.*

Proof. For each $n \in I^+$, let $D_n = \{x_j^n \mid j \in I^+ \cup \{0\}\}$ be a countable dense subset of S_n. Denote by F the set of all functions from finite subsets of I^+ into $I^+ \cup \{0\}$. For each $f \in F$ and $n \in I^+$, let

$$x_f(n) = \begin{cases} x^n_{f(n)} & \text{if } n \text{ is in the domain of } f \\ x^n_0 & \text{otherwise.} \end{cases}$$

Thus $x_f : I^+ \to \bigcup\{S_n \mid n \in I^+\}$, and so $D = \{x_f \mid f \in F\}$ is a countable subset of $\prod\{S_n \mid n \in I^+\}$. Let $B = \bigcap\{\pi_{n_j}^{-1}(G_{n_j}) \mid G_{n_j} \in \tau_{n_j}, j = 1, 2, \ldots, k\}$ be any member of the defining base for the product topology on $\prod\{S_n \mid n \in I^+\}$. For $j = 1, 2, \ldots, k$, let $m_j \in I^+ \cup \{0\}$ such that $x^{n_j}_{m_j} \in G_{n_j}$. Let $f(n_j) = m_j$ for $j = 1, 2, \ldots, k$. Thus $x_f \in D$ and $x_f \in B$, since $x_f(n_j) = x^{n_j}_{m_j}, j = 1, 2, \ldots, k$. Since $B \cap D \neq \varnothing$, D is dense in $\prod\{S_n \mid n \in I^+\}$.

THEOREM 1-17. *If $\langle S, \tau_1\rangle$ is separable and $f: \langle S, \tau_1\rangle \xrightarrow{\text{onto}} \langle T, \tau_2\rangle$ is continuous, then $\langle T, \tau_2\rangle$ is separable.*

Proof. Left as an exercise for the reader.

EXERCISES

1-30. Prove Theorem 1-15.

1-31. Prove Theorem 1-17.

1-7 First and Second Countable Spaces

In this section we examine two types of topological spaces: (1) those spaces $\langle S, \tau\rangle$ such that τ has a countable base, which we call "second countable"; and (2) those spaces $\langle S, \tau\rangle$ such that τ has a countable local base at p for every $p \in S$, which we call "first countable." Clearly, every second countable space is also first countable. Moreover, a second countable space is hereditarily separable, but a hereditarily separable space is not necessarily second countable. The two concepts are equivalent in a metric space that we define in the next section.

DEFINITION 1-24. $\langle S, \tau\rangle$ is *second countable* iff τ has a countable base $\mathscr{B} = \{B_n \mid n \in I^+\}$.

EXAMPLE 1-16. $\langle R, \xi\rangle$ is second countable. $\mathscr{B} = \{(r - 1/n, r + 1/n) \mid r$ is rational, $n \in I^+\}$ is a countable base for ξ. However, $\langle R, \mathscr{L}\rangle$ and the reals under the discrete topology are not second countable. They are both first countable.

DEFINITION 1-25. Let $\langle S, \tau\rangle$ be a topological space and $p \in S$. $\mathscr{B}_p \subset \tau$ is a *local base at* p iff $p \in G \in \tau$ implies that $\exists B \in \mathscr{B}_p$ such that $p \in B \subset G$.

DEFINITION 1-26. $\langle S, \tau\rangle$ is *first countable* iff there exists a countable base $\mathscr{B}_p$ at each $p \in S$.

THEOREM 1-18. *Every second countable space $\langle S, \tau\rangle$ is separable.*

Proof. Let $\mathscr{B} = \{B_n \mid n \in I^+\}$ be a countable base for τ. Let $x_n \in B_n$, $\forall n \in I^+$, and $D = \{x_n \mid n \in I^+\}$. D is a countable subset of S, which we show to be dense in S. Let $x \in S$ and $x \in G \in \tau$. Then $\exists B_n \in \mathscr{B}$ such that $x \in B_n \subset G$. Since $B_n \cap D \neq \varnothing$ implies that $G \cap D \neq \varnothing$, $x \in \bar{D}$ and D is dense in S.

THEOREM 1-19. *Let $\langle S, \tau\rangle$ be first countable, $A \subset S$, and $x \in S$. Then $x \in \bar{A}$ iff $\exists x_n \in A$ such that $x_n \to x$.*

Proof. If $x_n \in A$ and $x_n \to x$, then $x \in \bar{A}$ by Theorem 1-9(1). Suppose now that $x \in \bar{A}$. Let $x_n \in B_n$, $\forall n \in I^+$, where $\{B_n \mid n \in I^+\}$ is a countable local base at x with $B_{n+1} \subset B_n$, $\forall n \in I^+$. (Exercise 1-35 guarantees that such a "nested" local base at x exists.) If $x \in G \in \tau$, then by Definition 1-25 $\exists N \in I^+$ such that $x \in B_N \subset G$. This implies that $x_n \in G$, $\forall n \geq N$, and $x_n \to x$.

THEOREM 1-20. *If $\langle S, \tau_1\rangle$ is first countable, $f\colon \langle S, \tau_1\rangle \to \langle T, \tau_2\rangle$, and $f(x_n) \to f(x)$ in T whenever $x_n \to x$ in S, then f is continuous.*

Proof. Left as an exercise for the reader.

THEOREM 1-21. *If $\langle S_n, \tau_n\rangle$ is first countable $\forall n \in I^+$, then $\prod\{\langle S_n, \tau_n\rangle \mid n \in I^+\}$ is also first countable.*

Proof. Let $x \in \prod\{S_n \mid n \in I^+\}$ and $\{B_j^n \mid j \in I^+\}$ be a countable base at $\pi_n(x)$, $\forall n \in I^+$. Denote by $\mathscr{B}$ the collection of all intersections of finitely many sets of the form $\pi_n^{-1}(B_j^n)$, $n, j \in I^+$. Also, $\mathscr{B}$ is countable; its elements are members of the product topology, and each contains x. If $x \in B = \bigcap\{\pi_n^{-1}(G_n) \mid G_n \in \tau_n, n = 1, 2, \ldots, k\}$, then $\pi_n(x) \in G_n$ for $n = 1, 2, \ldots, k$. Hence $\exists j_n \in I^+$ such that $\pi_n(x) \in B_{j_n}^n \subset G_n$ for $n = 1, 2, \ldots, k$. This implies that $x \in \bigcap\{\pi_n^{-1}(B_{j_n}^n) \mid n = 1, 2, \ldots, k\} \in \mathscr{B}$, and $\mathscr{B}$ is a local base at x.

Since homeomorphisms preserve open sets, first and second countability are easily seen to be topological properties. However, they are not continuous invariants (as separability is), as shown by the following example.

EXAMPLE 1-17. $\langle R, \xi \rangle$ is second countable. $\langle R$, cofinite$\rangle$ is *not* first countable, since the complement of every finite set is open and every local base contains uncountably many such sets. The function $f: \langle R, \xi \rangle \to \langle R$, cofinite$\rangle$ given by $f(x) = x$ is continuous.

EXERCISES

1-32. Show that first and second countability are hereditary properties.

1-33. Show that every second countable space is hereditarily separable.

1-34. Prove that the product of countably many second countable spaces is second countable.

1-35. Let $\langle S, \tau \rangle$ be a topological space and $\mathscr{B}$ be a base for τ. Show that $\langle S, \tau \rangle$ is first countable iff $\forall x \in S$ there exists a "nested" sequence $\{B_n\}_{n \in I^+}$ of elements of $\mathscr{B}$ containing x such that $x \in G \in \tau$ implies that $\exists N \in I^+$ such that $B_n \subset G$, $\forall n \geq N$. "Nested" means that $B_{n+1} \subset B_n$, $\forall n \in I^+$.

1-36. Prove Theorem 1-20.

1-8 Distance, Uniform, and Proximity Spaces

In Section 1-1 we discussed the topologizing of a set by specifying the "open sets" or by specifying the "closed sets" or by using either the closure operator or the interior operator. In this section we discuss three additional ways of topologizing a set: (1) using a distance function, (2) using a uniformity, and (3) using a proximity. These three methods are related, because (2) and (3) may be regarded as generalizations of (1).

DEFINITION 1-27. Let $S \neq \varnothing$ and $\rho : S \times S \to R^+ \cup \{0\}$ be a function such that the following conditions are satisfied:

(M1) $\rho(x, y) = 0$ iff $x = y$, $\forall x, y \in S$.

(M2) $\rho(x, y) = \rho(y, x)$, $\forall x, y \in S$ (i.e., ρ is symmetric).

(M3) $\rho(x, y) \leq \rho(x, z) + \rho(z, y)$, $\forall x, y, z \in S$ (i.e., ρ satisfies the triangle inequality).

Then ρ is a *metric* on S, and $\langle S, \rho \rangle$ is a *metric set*.

If ρ satisfies (M1) and (M2), ρ is a *semimetric* on S.

If ρ satisfies (M1) and (M3), ρ is a *quasimetric* on S.

If ρ satisfies (M2), (M3), and $\rho(x, x) = 0$, $\forall x \in S$, then ρ is a *pseudometric* on S.

We show that every metric on S induces a topology on S called the "metric topology." Thus every metric set is a metric space. We also show by example that a semimetric on a set does not necessarily induce a topology on the set.

THEOREM 1-22. *If $\langle S, \rho \rangle$ is a metric set, then ρ induces a topology on S having $\mathscr{B} = \{S_\rho(p; r) \mid p \in S, r > 0\}$ as a base, where $S_\rho(p; r) = \{x \in S \mid \rho(p, x) < r\}$.*

Proof. If $p \in S$, then $p \in S_\rho(p; r)$, $\forall r > 0$, and condition (1) of Theorem 1-5 is satisfied. Now let $x \in S_\rho(p_1; r_1) \cap S_\rho(p_2; r_2)$ and let $r = \min\{r_1 - \rho(p_1, x), r_2 - \rho(p_2, x)\}$. If $y \in S_\rho(x; r)$, then $\rho(x, y) < r \leq r_1 - \rho(p_1, x)$ and $\rho(x, y) < r \leq r_2 - \rho(p_2, x)$. Thus, by the triangle inequality, $\rho(p_1, y) < r_1$ and $\rho(p_2, y) < r_2$, which implies that $y \in S_\rho(p_1; r_1) \cap S_\rho(p_2; r_2)$. Hence $S_\rho(x; r) \subset S_\rho(p_1; r_1) \cap S_\rho(p_2, r_2)$, and condition (2) is satisfied. Thus $\mathscr{B}$ is the base for a topology on S.

EXAMPLE 1-18. Let $\rho(x, y) = |x - y|$, $\forall x, y \in R$. Clearly, $|x - y| = 0$ iff $x - y = 0$ iff $x = y$ and $|y - x| = |x - y|$, so that (M1) and (M2) are satisfied. Also, $x - y = (x - z) + (z - y)$ implies that $|x - y| = |(x - z) + (z - y)| \leq |x - z| + |z - y|$, and (M3) is satisfied. Thus $\langle R, \rho \rangle$ is a metric space. We leave as an exercise for the reader the verification of the fact that the ρ-metric topology on R is ξ.

EXAMPLE 1-19. Let $\rho(x, y) = |x - y|$ if x and y are both rational or both irrational; let $\rho(x, y) = |x - y|^{-1}$ otherwise. Clearly, $\rho(x, y) = 0$ iff $x = y$ and $\rho(x, y) = \rho(y, x)$. Hence ρ is a semimetric on R; ρ is not a metric since (M3) is not satisfied. For let $x = 5$, $y = 6$, $z = \sqrt{2}$, $\rho(5, \sqrt{2}) + \rho(\sqrt{2}, 6) = (5 - \sqrt{2})^{-1} + (6 - \sqrt{2})^{-1} = (11 - 2\sqrt{2})(32 - 11\sqrt{2})^{-1} < 1 = \rho(5, 6)$. Let $0 < \varepsilon < 1$. $S_\rho(0; \varepsilon)$ consists of all rationals in $(-\varepsilon, \varepsilon)$ and all irrationals in $(-\infty, -1/\varepsilon) \cup (1/\varepsilon, \infty)$. If x is any irrational point in $S_\rho(0; \varepsilon)$, there does *not* exist any spherical neighborhood $S_\rho(x; \delta)$ about x that is contained in $S_\rho(0; \varepsilon)$. Hence $\mathscr{B} = \{S_\rho(x; r) \mid x \in R, r > 0\}$ is not a base for any topology on R.

Next we investigate the concept of a "uniform space," which is a generalization of a metric space. Excellent treatments of uniform spaces are contained in Kelley's *General Topology* and Pervin's *Foundations on General Topology* (see the Bibliography).

DEFINITION 1-28. Let $S \neq \varnothing$ and $\mathscr{U} \subset 2^{S \times S}$, satisfying the following axioms:

(U1) $\Delta = \{\langle x, x \rangle \mid x \in S\} \subset U$, $\forall U \in \mathscr{U}$.

(U2) If $U \in \mathcal{U}$ and $U \subset V$, then $V \in \mathcal{U}$.

(U3) If $U, V \in \mathcal{U}$, then $U \cap V \in \mathcal{U}$.

(U4) $\forall U \in \mathcal{U}\ \exists V \in \mathcal{U}$ such that $V \circ V \subset U$.

(U5) $U \in \mathcal{U}$ implies that $U^{-1} \in \mathcal{U}$.

Then $\mathcal{U}$ is called a *uniformity for S* and $\langle S, \mathcal{U} \rangle$ is a *uniform space*. If $\mathcal{U}$ satisfies (U1)–(U4), then $\mathcal{U}$ is a *quasiuniformity for S*. In either case, a base for $\mathcal{U}$ is any subcollection $\mathcal{B}$ of $\mathcal{U}$ with the property that each member of $\mathcal{U}$ contains a member of $\mathcal{B}$.

Conditions (U1), (U4), and (U5) on a uniformity correspond roughly to conditions (M1), (M3), and (M2), respectively, on a metric. In what follows, we show that every metric space is a uniform space and define the "uniform topology $\tau_{\mathcal{U}}$," which is induced on a set S by a uniformity $\mathcal{U}$ for S.

THEOREM 1-23. *Every metric space $\langle M, \rho \rangle$ is a uniform space.*

Proof. Let $B_\varepsilon = \{\langle x, y \rangle \in M \times M \mid \rho(x, y) < \varepsilon\}$ for each $\varepsilon > 0$ and let $\mathcal{U} = \{U \subset M \times M \mid U \supset B_\varepsilon \text{ for some } \varepsilon > 0\}$. $\Delta \subset U\ \forall U \in \mathcal{U}$, since $\Delta \subset B_\varepsilon$ $\forall \varepsilon > 0$ and (U1) is satisfied. If $U \in \mathcal{U}$ and $U \subset V$, then $B_\varepsilon \subset U \subset V$ for some $\varepsilon > 0$. Hence $V \in \mathcal{U}$ and (U2) is satisfied. If $U, V \in \mathcal{U}$, then $B_{\varepsilon_1} \subset U$ and $B_{\varepsilon_2} \subset V$ for some $\varepsilon_1, \varepsilon_2 > 0$. If $\varepsilon_1 \geq \varepsilon_2$, then $B_{\varepsilon_2} \subset B_{\varepsilon_1} \subset U$ and $B_{\varepsilon_2} \subset V$. Hence $B_{\varepsilon_2} \subset U \cap V$ and $U \cap V \in \mathcal{U}$. If $\varepsilon_1 < \varepsilon_2$, then $B_{\varepsilon_1} \subset B_{\varepsilon_2} \subset V$ and $B_{\varepsilon_1} \subset U$. Hence $B_{\varepsilon_1} \subset U \cap V$ and $U \cap V \in \mathcal{U}$. Thus (U3) is satisfied. To see that (U4) is satisfied, we let $B_\varepsilon \subset U \in \mathcal{U}$ and $V = B_{\varepsilon/2}$. Then $B_{\varepsilon/2} \circ B_{\varepsilon/2} = \{\langle x, y \rangle \in M \times M \mid \exists z \in M \text{ such that } \rho(x, z) < \varepsilon/2 \text{ and } \rho(z, y) < \varepsilon/2\} \subset B_\varepsilon \subset U$. Finally, since ρ is symmetric, if $B_\varepsilon \subset U \in \mathcal{U}$, then $B_\varepsilon \subset U^{-1}$ also. Thus $U \in \mathcal{U}$ implies that $U^{-1} \in \mathcal{U}$ and (U5) is satisfied. By Definition 1-28, $\mathcal{U}$ is a uniformity for M.

Sometimes it is useful to describe the structure on a topological space $\langle S, \tau \rangle$ in terms of the "neighborhood systems" of the points of S. We say that $A \subset S$ is a *neighborhood* of $x \in S$ if and only if $\exists G \in \tau$ such that $x \in G \subset A$. The family $\mathcal{N}_x$ of all such neighborhoods of x is called the *neighborhood system* of x. A *neighborhood base* for $\mathcal{N}_x$ is any subcollection $\mathcal{B}_x$ of $\mathcal{N}_x$ such that each $A \in \mathcal{N}_x$ contains some $B \in \mathcal{B}_x$.

DEFINITION 1-29. Let $\langle S, \mathcal{U} \rangle$ be a uniform space. For each $x \in S$ and each $U \in \mathcal{U}$, let $U[x] = \{y \in S \mid \langle x, y \rangle \in U\}$. The collection $\mathcal{B}_x = \{U[x] \mid U \in \mathcal{U}\}$ for each $x \in S$ is a neighborhood base for a topology $\tau_{\mathcal{U}}$ on S called the *uniform topology* on S induced by the uniformity $\mathcal{U}$.

REMARK. It should be noted that two different metrics on a set S can yield the same uniformity for S by means of the procedure outlined in the proof of Theorem 1-23. Also, two different uniformities on a set S can produce the same uniform topology by Definition 1-29. For example, the metrics $d(x, y) = |x - y|$ and $\rho(x, y) = 2|x - y|$ produce the same uniformity on the set R of reals. Moreover, the discrete uniformity and the uniformity having as a base $\mathscr{B} = \{\{\Delta \cup \{\langle x, y\rangle \mid x > r, y > r\}\} \mid r \in R\}$ both induce the discrete topology on R.

EXAMPLE 1-20. Let R be the set of reals and $\rho(x, y) = \max\{x - y, 0\}$ $\forall\langle x, y\rangle \in R \times R$. Clearly, $\rho(x, x) = 0$ for each $x \in R$ and ρ satisfies (M3) on R. Let $B_\varepsilon = \{\langle x, y\rangle \in R \times R \mid \rho(x, y) < \varepsilon\}$ for each $\varepsilon > 0$. The collection $\mathscr{B} = \{B_\varepsilon \mid \varepsilon > 0\}$ is a base for a quasiuniformity $\mathscr{U}$ on R in like manner, as it was a base for a uniformity on S in the metric case in Theorem 1-23. $\mathscr{U}$ is not a uniformity, because (U5) is not satisfied, since ρ does not satisfy (M2). Since $B_\varepsilon = \{\langle x, y\rangle \in R \times R \mid y > x - \varepsilon\}$ for each $\varepsilon > 0$, the reader may easily verify that the quasiuniform topology induced on R by $\mathscr{U}$ is $\tau_{\mathscr{U}} = \{\varnothing\} \cup \{R\} \cup \{(a, +\infty) \mid a \in R\}$. We encountered this topology for the reals earlier, in Exercise 1-12.

We conclude this section with a brief mention of proximity spaces, another generalization of metric spaces. The topological spaces derivable from proximity spaces are precisely those which are uniformizable. Additional information about proximity spaces can be found in Pervin's *Foundations of General Topology* and in material referenced by Pervin.

DEFINITION 1-30. $\langle S, \delta\rangle$ is a *proximity space* iff $S \neq \varnothing$ and δ is a relation on 2^S satisfying the following conditions:

(P1) $\langle A, \varnothing\rangle \notin \delta$, $\forall A \in 2^S$.

(P2) $\langle\{x\}, \{x\}\rangle \in \delta$, $\forall x \in S$.

(P3) $\langle C, A \cup B\rangle \in \delta$ iff $\langle C, A\rangle \in \delta$ or $\langle C, B\rangle \in \delta$ $\forall A, B, C \in 2^S$.

(P4) If $\langle A, B\rangle \notin \delta$, then $\exists C \in 2^S$ such that $\langle A, C\rangle \notin \delta$ and $\langle S - C, B\rangle \notin \delta$.

(P5) $\langle A, B\rangle \in \delta$ iff $\langle B, A\rangle \in \delta$.

The relation δ is called a *proximity* for S, and "$\langle A, B\rangle \in \delta$" is read "$A$ is near B."

EXAMPLE 1-21. Let $\langle S, \rho\rangle$ be a pseudometric space. If $A, B \in 2^S$, let $\langle A, B\rangle \in \delta$ iff $\rho(A, B) \equiv \inf\{\rho(x, y) \mid x \in A, y \in B\} = 0$. $\rho(A, \varnothing) = \infty$ implies that $\langle A, \varnothing\rangle \notin \delta$, $\forall A \in 2^S$, and (P1) is satisfied. Since $\rho(x, x) = 0$, $\langle\{x\}, \{x\}\rangle \in \delta$, $\forall x \in S$, and (P2) is satisfied. Since $\rho(C, A \cup B) \leq \rho(C, A)$ and $\rho(C, A \cup B) \leq \rho(C, B)$, we have $\rho(C, A \cup B) = 0$ iff $\rho(C, A) = 0$ or $\rho(C, B) = 0$. Thus $\langle C, A \cup B\rangle \in \delta$ iff $\langle C, A\rangle \in \delta$ or $\langle C, B\rangle \in \delta$, and (P3) is satisfied. If $\langle A, B\rangle \notin \delta$, then $\rho(A, B) = r > 0$. Let $C = \{x \in S \mid \rho(\{x\}, B) \leq r/2\}$. Thus $\rho(A, C) \geq$

$r/2 > 0$ and $\rho(S - C, B) \geq r/2 > 0$, which imply that $\langle A, C\rangle \notin \delta$ and $\langle S - C, B\rangle \notin \delta$. Hence (P4) is satisfied. Finally, $\rho(A, B) = \rho(B, A)$ implies that $\langle A, B\rangle \in \delta$ iff $\langle B, A\rangle \in \delta$, and (P5) is satisfied. Thus ρ is a proximity for S.

THEOREM 1-24. *Every proximity space $\langle S, \delta\rangle$ is a topological space.*

Proof. Let $c(A) = \{x \in S \mid \langle\{x\}, A\rangle \in \delta\}$ for each $A \in 2^S$. It is easily checked that c satisfies (K1)–(K4) of Theorem 1-4 and is a closure operator on 2^S. The closure topology is obtained by defining $A \in 2^S$ to be closed iff $c(A) = A$ and $A \in 2^S$ to be open (i.e., a member of the topology) iff $S - A$ is closed.

THEOREM 1-25. *The topological spaces derivable from the proximity spaces are precisely the uniformizable spaces.*

Proof. Let $\langle S, \mathscr{U}\rangle$ be a uniform space. If $A, B \in 2^S$, we let $\langle A, B\rangle \in \delta$ iff $\forall U \in \mathscr{U}\ \exists x \in A, y \in B$ such that $\langle x, y\rangle \in U$. Then δ is a proximity for S. On the other hand, if $\langle S, \delta\rangle$ is a proximity space, we let $U(A, B) = S \times S - [(A \times B) \cup (B \times A)]$, $\forall A, B \in 2^S$. Then the collection $\mathscr{S} = \{U(A, B) \mid \langle A, B\rangle \notin \delta\}$ is a subbase for a uniformity for S compatible with δ.

EXERCISES

1-37. Let $\mathscr{C}[0, 1]$ be the collection of all continuous, real-valued functions defined on $I^1 = [0, 1]$. Determine whether or not each of the following is a metric on $\mathscr{C}[0, 1]$:

(a) $\rho(f, g) = \sup\{|f(x) - g(x)| \mid x \in I^1\}$.

(b) $\rho(f, g) = \int_0^1 |f(x) - g(x)|\,dx$.

1-38. If $\langle S, \rho\rangle$ is a metric space, show that $\rho : S \times S \to R^+ \cup \{0\}$ is continuous.

1-39. Let $\langle S, \rho\rangle$ be a metric space. We define $\rho(A, B) = \inf\{\rho(x, y) \mid x \in A, y \in B\}$ for $A, B \in 2^S$. Show that $x \in \bar{A}$ iff $\rho(\{x\}, A) = 0$.

1-40. Show that every metric space is first countable. Moreover, show that a metric space is second countable iff separable.

1-41. If $\langle S, \rho\rangle$ is a metric space, show that $x_n \to x$ in S iff $\rho(x_n, x) \to 0$.

1-42. Let $\langle S, \rho\rangle$ be a metric space and $\varnothing \neq A \subset S$. We define the *diameter* of A to be $\delta(A) = \sup\{\rho(x, y) \mid x, y \in A\}$ and say that A is *bounded* iff

$\delta(A) < \infty$. Let $\rho^*(x, y) = \rho(x, y)(1 + \rho(x, y))^{-1}$, $\forall x, y \in S$. Show that ρ^* is a bounded metric for S and the ρ^*-metric topology is identical with the ρ-metric topology on S.

1-43. Let $\langle S, \tau \rangle$ be a topological space and $A \subset S$. A is a "G_δ" iff A is expressible as the intersection of a countable number of open sets (i.e., members of τ). A is an "F_σ" iff A is expressible as the union of a countable number of closed sets. If $\langle S, \rho \rangle$ is a metric space, show that each closed subspace is a G_δ and each open subspace is an F_σ.

1-44. Let $\langle S, \mathscr{U} \rangle$ be a uniform space and $U \in \mathscr{U}$. Show that there exists a symmetric set $V \in \mathscr{U}$ such that $V \circ V \subset U$.

1-45. Let $\langle S, \delta \rangle$ be a proximity space and $A, B \in 2^S$. Show that $\langle A, B \rangle \in \delta$ iff $\langle \bar{A}, \bar{B} \rangle \in \delta$.

★1-9 Function Spaces and Quotient Spaces

We conclude Chapter 1 with a brief discussion of function spaces and quotient spaces. For collections of functions from one set into another, we define two useful topologies:

(1) The topology of pointwise convergence (or point-open topology).
(2) The topology of uniform convergence.

In Chapter 3 we shall define the topology of compact convergence (or compact-open topology).

DEFINITION 1-31. Let $S \neq \varnothing$ be a set and $\langle T, \tau_1 \rangle$ a topological space. Denote by $\mathscr{F}(S, T)$ the collection of all functions from S into T. The collection $\mathscr{S} = \{\{f \in \mathscr{F}(S, T) \mid f(x_0) \in G\} \mid x_0 \in S, G \in \tau_1\}$ is a subbase for the *topology of pointwise convergence* (or *point-open topology*) τ_2 on $\mathscr{F}(S, T)$.

EXAMPLE 1-22. Let $S = [0, 1]$, $\langle T, \tau_1 \rangle = \langle R, \xi \rangle$ and τ_2 be the point open topology on $\mathscr{F}(S, T)$. Let $f_n(x) = x^n$, $\forall n \in I^+$. Then $\{f_n\}_{n \in I^+}$ converges pointwise to the function $g : [0, 1] \to R$ given by

$$g(x) = \begin{cases} 0 & \text{if } 0 \leq x < 1 \\ 1 & \text{if } x = 1. \end{cases}$$

See Figure 1-2. Note that g is *not* continuous even though each f_n is. This motivates us to define the topology of uniform convergence, since the uniform limit of a sequence of continuous functions is continuous.

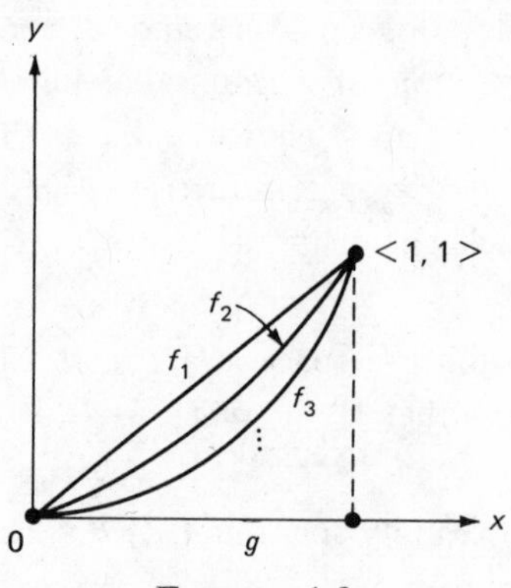

FIGURE 1-2

DEFINITION 1-32. Let $S \neq \varnothing$ be a set and $\langle T, d\rangle$ a metric space. Denote by $\mathscr{B}(S, T)$ the collection of bounded functions from S into T. If $f, g \in \mathscr{B}(S, T)$, let $\rho(f, g) = \sup\{d(f(x), g(x)) \mid x \in S\}$. Then ρ is a metric that induces on $\mathscr{B}(S, T)$ the *topology of uniform convergence*, having as a base the collection $\{S_\rho(f; \varepsilon) \mid f \in \mathscr{B}(S, T), \varepsilon > 0\}$, where $S_\rho(f; \varepsilon) = \{g \in \mathscr{B}(S, T) \mid \rho(f, g) < \varepsilon\}$.

EXAMPLE 1-23. Let $S = T = [0, 1]$ and $d(x, y) = |x - y| \ \forall x, y \in T$. Also let $\mathscr{B}(S, T)$ be endowed with the topology of uniform convergence induced by the metric ρ, defined by $\rho(f, g) = \sup\{|f(x) - g(x)| \mid x \in [0, 1]\}$ as described in Definition 1-32. Consider the sequence of bounded continuous functions $f_n : S \to T$, given by

$$f_n(x) = \begin{cases} x & \text{if } 0 \leq x \leq \dfrac{n}{n+1} \\ \dfrac{n}{n+1} & \text{if } \dfrac{n}{n+1} \leq x \leq 1 \end{cases}$$

for each $n \in I^+$. Let $g : S \to T$ be the identity function defined by $g(x) = x$, $\forall x \in S$. Their graphs are shown in Figure 1-3. Since

$$\rho(f_n, g) = \sup\left\{\left|\frac{n}{n+1} - x\right| \;\middle|\; \frac{n}{n+1} \leq x \leq 1\right\} \leq 1 - \frac{n}{n+1} = \frac{1}{n+1},$$

we see that $\{f_n\}_{n \in I^+}$ converges uniformly to the bounded continuous function g.

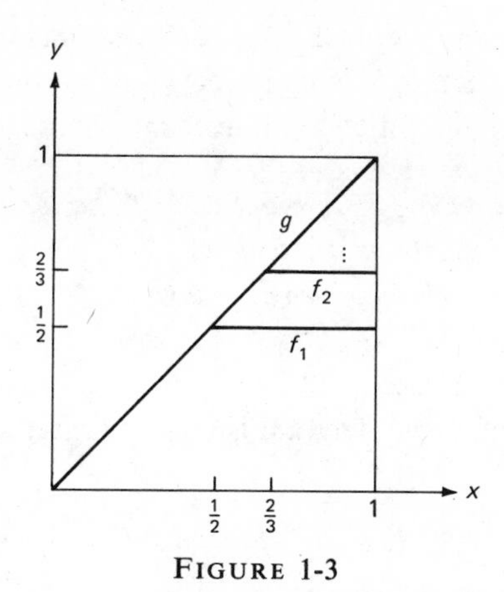

FIGURE 1-3

Finally, we introduce the concept of "quotient topology." Moreover, an equivalence relation R on a set S determines a quotient set S/R, which is called a "quotient space" when endowed with the quotient topology.

DEFINITION 1-33. Let $\langle S, \tau \rangle$ be a topological space, $T \neq \varnothing$ be a set, and $f(S) = T$. The *quotient topology* $\mathscr{G}$ is the largest topology on T such that f is continuous.

DEFINITION 1-34. Let $S \neq \varnothing$ be a set and R an equivalence relation on S. Denote by S/R the collection of R-equivalence classes in S and by φ the canonical mapping of S onto S/R given by $\varphi(x) = [x]$. If $\mathscr{G}$ is the quotient topology on S/R determined by φ, then $\langle S/R, \mathscr{G} \rangle$ is a *quotient space.*

EXAMPLE 1-24. Let $S = [0, 1]$ and suppose that $\langle x, y \rangle \in R$ iff x and y are both rational or both irrational. Then R is an equivalence relation on S, and the quotient space S/R is the two-point indiscrete space.

EXAMPLE 1-25. Let $S = [0, 1]$ and $R = \{\langle x, x \rangle \mid x \in S\} \cup \{\langle 0, 1 \rangle, \langle 1, 0 \rangle\}$. Then R is an equivalence relation on S which identifies the two endpoints. Thus the quotient space S/R is homeomorphic to the unit circle S^1. To see this, let $\varphi : S \to S/R$ be the canonical mapping of S onto S/R. Since S/R has the quotient topology, φ is continuous. If $f\colon [0, 1] \to S^1$ denotes the continuous function defined by $f(x) = e^{2\pi i x}$, $\forall x \in [0, 1]$, then the reader can easily verify that the function $f \circ \varphi^{-1} : S/R \to S^1$ is bijective, continuous, and open (hence a homeomorphism).

As remarked in Example 1-25, the canonical mapping $\varphi : S \to S/R$ is always continuous, since S/R is endowed with the quotient topology. We use this fact to prove the following interesting theorem.

THEOREM 1-26. *Let $\langle S, \tau_1 \rangle$ and $\langle T, \tau_2 \rangle$ be topological spaces having respective equivalence relations R_1 and R_2. If $f : \langle S, \tau_1 \rangle \to \langle T, \tau_2 \rangle$ is continuous and relation-preserving, then the function $f_* : S/R_1 \to T/R_2$ given by $f_*([x]) = [f(x)]$ (where $[x] = \{x' \in S \mid \langle x, x' \rangle \in R_1\}$ and $[f(x)] = \{y' \in T \mid \langle f(x), y' \rangle \in R_2\}$) is continuous.*

Proof. Since the following diagram is commutative,

$$\begin{array}{ccc} S & \xrightarrow{f} & T \\ {\scriptstyle \varphi_1}\downarrow & & \downarrow{\scriptstyle \varphi_2} \\ S/R_1 & \xrightarrow[f_*]{} & T/R_2 \end{array}$$

we have that $\varphi_2 \circ f = f_* \circ \varphi_1$. Moreover, the continuity of $\varphi_2 \circ f$ and φ_2 implies the continuity of $f_* \circ \varphi_1$ and f_*.

By identifying in a pseudometric space all those points at zero distance apart, we get a quotient space that is metrizable.

THEOREM 1-27. *Let $\langle S, \rho \rangle$ be a pseudometric space and $\langle x, y \rangle \in R$ iff $\rho(x, y) = 0$, $\forall x, y \in S$. Then R is an equivalence relation on S and $\langle S/R, \mathscr{G} \rangle$ is metrizable.*

Proof. $\langle x, x \rangle \in R$, since $\rho(x, x) = 0$, $\forall x \in S$. Also, $\langle x, y \rangle \in R$ implies that $\langle y, x \rangle \in R$, since $\rho(x, y) = \rho(y, x)$, $\forall x, y \in S$. Finally, if $\langle x, y \rangle \in R$ and $\langle y, z \rangle \in R$, then $\rho(x, y) = \rho(y, x) = 0$. Hence $0 \leq \rho(x, z) \leq \rho(x, y) + \rho(y, z) = 0$ implies that $\rho(x, z) = 0$ and $\langle x, z \rangle \in R$. Thus R is an equivalence relation on S. Let $\rho^*([x], [y]) = \rho(x, y)$, where $x \in [x]$, $y \in [y]$, $\forall [x]$,

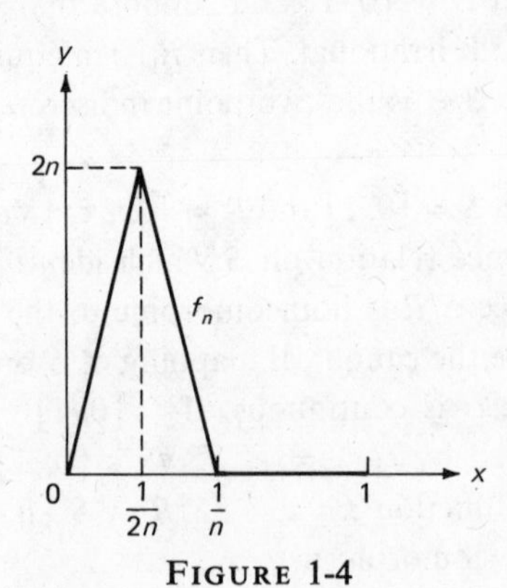

FIGURE 1-4

$[y] \in S/R$. Then ρ^* is a metric on S/R. Moreover, the ρ^*-metric topology on S/R coincides with the quotient topology $\mathcal{G}$.

EXERCISES

1-46. Let $\mathcal{F}(I^1, R)$ have the topology of pointwise convergence and define $\forall n \in I^+$:

$$f_n(x) = \begin{cases} 4n^2x & \text{if } 0 \le x \le 1/2n \\ -4n^2x + 4n & \text{if } 1/2n < x < 1/n \\ 0 & \text{if } 1/n \le x \le 1. \end{cases}$$

Show that $\{f_n\}_{n \in I^+}$ converges pointwise to $g(x) \equiv 0$ on I^1.

1-47. Let $\langle S, \tau_1 \rangle$ and $\langle T, \tau_2 \rangle$ be topological spaces. If $\{f_n\}_{n \in I^+}$ is a sequence of functions in $\mathcal{F}(S, T)$ with the topology of pointwise convergence, then show that $\lim_{n \to \infty} f_n = g$ iff $\lim_{n \to \infty} f_n(x) = g(x)$, $\forall x \in S$.

1-48. Show that if $f: S \to T$ and $G \subset T$, then $G \in \mathcal{G}$ (the quotient topology on T) iff $f^{-1}(G) \in \tau$ (the given topology on S). Also, $C \subset T$ is closed iff $f^{-1}(C)$ is closed.

1-49. If $f: \langle S, \tau_1 \rangle \xrightarrow{\text{onto}} \langle T, \tau_2 \rangle$ is continuous and either open or closed, show that τ_2 is necessarily the quotient topology for T.

1-50. Prove that a mapping $f: S/R \to T$ is continuous iff $f \circ \varphi : S \to T$ is continuous.

1-51. Show that if A is a quotient space of S and B is a quotient space of A, then B is homeomorphic to a quotient space of S.

CHAPTER 2

The Separation Axioms

2-1 T_0, T_1, T_2, and $T_{5/2}$

Generally speaking, the topological properties of a space $\langle S, \tau \rangle$ depend primarily on τ, the collection of "open sets"; e.g., the cardinality of τ needs to be reasonably small for $\langle S, \tau \rangle$ to be separable or first or second countable. On the other hand, $f: \langle S, \tau_1 \rangle \to \langle T, \tau_2 \rangle$ is more likely to be continuous if the cardinality of τ_1 is large. In this chapter we discuss the T_i separation axioms of Alexandroff and Hopf, which deal with the distribution of the open sets in S. A metric space satisfies all the T_i-axioms. In this section, we consider those axioms dealing with the separation of two distinct points by open sets.

DEFINITION 2-1. $\langle S, \tau \rangle$ is a T_0-*space* iff $x, y \in S$ with $x \neq y$ implies that $\exists U \in \tau$ such that either $x \in U$ and $y \in S - U$, or $y \in U$ and $x \in S - U$. See Figure 2-1.

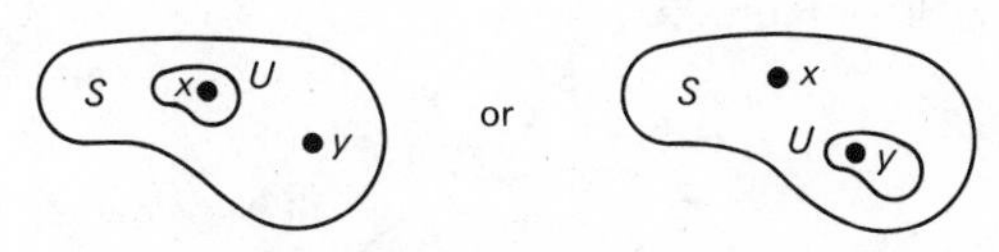

FIGURE 2-1

We now give an example of two topologies on $S = \{a, b, c\}$. They yield two spaces, one of which is a T_0-space. Since the T_0-axiom is the weakest separation property that we shall discuss, we may conclude that some topological spaces have "too few" open sets to be really useful.

EXAMPLE 2-1. Let $S = \{a, b, c\}$ and $\tau_1 = \{\varnothing, \{a\}, \{b\}, \{a, b\}, S\}$. Then, $\langle S, \tau_1 \rangle$ is a T_0-space, since $c \notin \{a\}$ and $c \notin \{b\}$. Let $\tau_2 = \{\varnothing, S\}$. $\langle S, \tau_2 \rangle$ is not a T_0-space, since $a, b, c \in S$.

DEFINITION 2-2. $\langle S, \tau \rangle$ is a T_1-*space* iff $x, y \in S$ with $x \neq y$ implies that $\exists U, V \in \tau$ with $x \in U$, $y \in S - U$ and $y \in V$, $x \in S - V$. See Figure 2-2.

The space $\langle S, \tau_1 \rangle$ described in Example 2-1 is a T_0-space that is not a T_1-space, because each open set containing c also contains a and b. If we

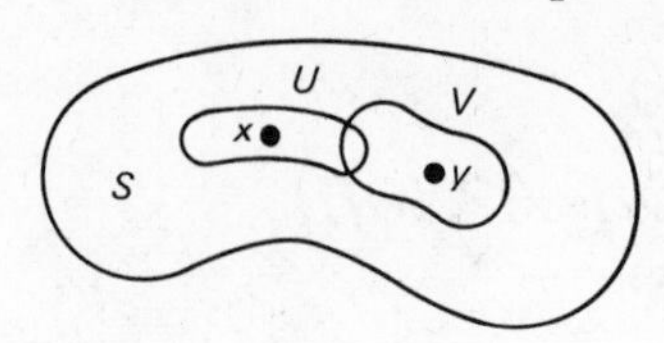

FIGURE 2-2

give S the discrete topology, however, then we get a T_1-space. As an exercise, the reader should verify that each T_1-space is a T_0-space. Definition 2-2 is equivalent to the requirement that all singletons be closed.

THEOREM 2-1. *$\langle S, \tau \rangle$ is a T_1-space iff $\overline{\{x\}} = \{x\}$, $\forall x \in S$.*

Proof. Let $\langle S, \tau \rangle$ be a T_1-space and $x \in S$. If $y \in S - \{x\}$, then $\exists V \in \tau$ such that $y \in V$ and $x \in S - V$. Hence $y \notin \overline{\{x\}}$ and $\overline{\{x\}} = \{x\}$. Conversely, suppose that $\overline{\{x\}} = \{x\}$, $\forall x \in S$. Let $y, z \in S$ with $y \neq z$. Then $\overline{\{y\}} = \{y\}$ implies that $\exists V \in \tau$ such that $z \in V$ and $y \in S - V$. Also, $\overline{\{z\}} = \{z\}$ implies that $\exists U \in \tau$ such that $y \in U$ and $z \in S - U$. Thus $\langle S, \tau \rangle$ is a T_1-space by Definition 2-2.

DEFINITION 2-3. $\langle S, \tau \rangle$ is a *T_2-space* iff $x, y \in S$ with $x \neq y$ implies that $\exists U, V \in \tau$ with $x \in U$, $y \in V$, and $U \cap V = \varnothing$. T_2-spaces are also called *Hausdorff spaces.* See Figure 2-3.

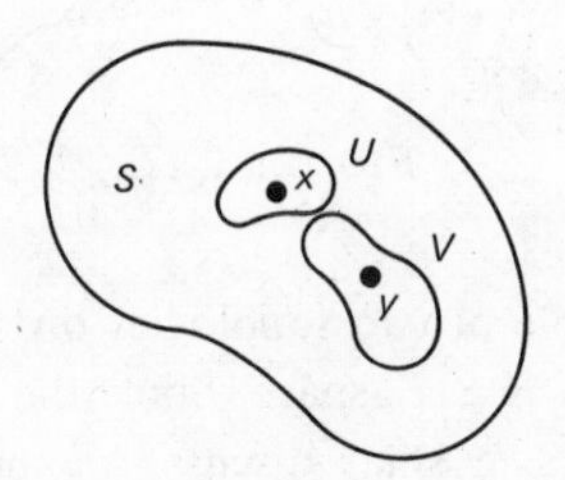

FIGURE 2-3

DEFINITION 2-4. $\langle S, \tau \rangle$ is a *$T_{5/2}$-space* iff $x, y \in S$ with $x \neq y$ implies that $\exists U, V \in \tau$ with $x \in U$, $y \in V$, and $\overline{U} \cap \overline{V} = \varnothing$. See Figure 2-4. (W. J. Thron calls such spaces *Urysohn spaces*, but "Urysohn spaces" for Steen and Seebach are more restrictive.)

The reader can easily check that an infinite set S together with the "cofinite topology" is a T_1-space that is not Hausdorff. We describe now a T_2-space that is not a Urysohn space. The example is due to Bing. We also give an example of a $T_{5/2}$-space due to Moore.

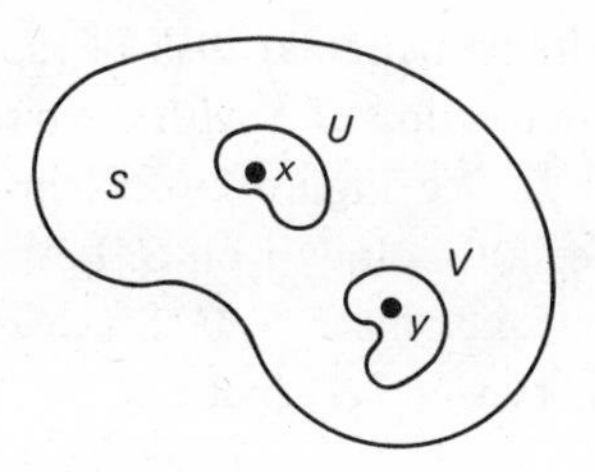

FIGURE 2-4

EXAMPLE 2-2. Let $S = \{\langle x, y\rangle \mid x, y \text{ are rational}, y \geq 0\}$. If $\langle a, b\rangle \in S$ and $\varepsilon > 0$, the set $\{\langle r, 0\rangle \mid \text{either } |r - (a + b/\sqrt{3})| < \varepsilon \text{ or } |r - (a - b/\sqrt{3})| < \varepsilon\} \cup \{\langle a, b\rangle\}$ is an ε-neighborhood of $\langle a, b\rangle$. Geometrically, such a neighborhood consists of $\langle a, b\rangle$ and all rational points in the intervals $(a - b/\sqrt{3} - \varepsilon, a - b/\sqrt{3} + \varepsilon)$ and $(a + b/\sqrt{3} - \varepsilon, a + b/\sqrt{3} + \varepsilon)$ centered about the feet of the equilateral triangle with vertex at $\langle a, b\rangle$ and base along the x-axis if $b > 0$. If $b = 0$, the ε-neighborhood of $\langle a, 0\rangle$ consists of all rational points in the interval $(a - \varepsilon, a + \varepsilon)$. See Figure 2-5. The collection of all such neighborhoods is a base for a T_2-topology τ on S which has the property that the intersection of the closures of any two members of $\tau - \{\varnothing\}$ is nonempty. Thus $\langle S, \tau\rangle$ is not a $T_{5/2}$-space.

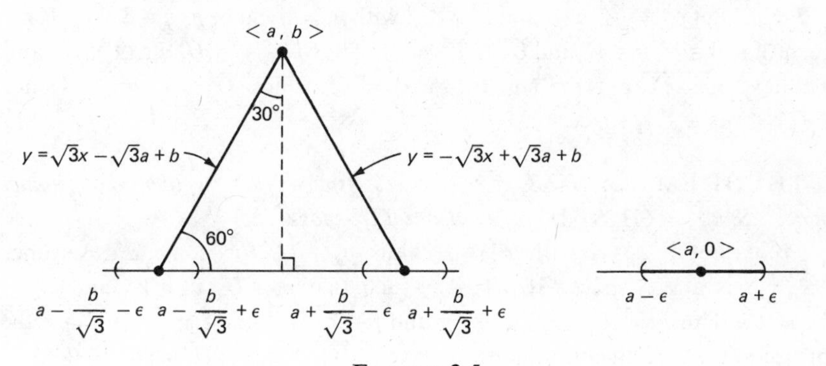

FIGURE 2-5

EXAMPLE 2-3. Let $S = \{\langle x, y\rangle \mid x, y \text{ are real}, y \geq 0\}$ and $L = \{\langle x, 0\rangle \mid x \text{ is real}\}$. Let d be the usual metric for E^2 and $S_d(p; r)$ the open d-sphere about p of radius $r > 0$. For each $p \in S$ and $r > 0$, we define a neighborhood $N_r(p)$ as follows:

(1) $N_r(p) = S_d(p; r) \cap S$ if $p \in S - L$.
(2) $N_r(p) = [S_d(p; r) \cap (S - L)] \cup \{p\}$ if $p \in L$.

Geometrically, $N_r(p)$ is the semicircular disk of radius r and centered at p if $p \in L$, and $N_r(p)$ is the intersection of S with the circular disk of radius r and centered at p if $p \in S - L$. See Figure 2-6. Clearly, the collection $\{N_r(p) \mid p \in S, r > 0\}$ is a base for a topology τ on S. Let $p, q \in S$ with $p \neq q$. Then $d(p, q) = |p - q| = 3r$ for some $r > 0$. Since $r = \frac{1}{3}d(p, q)$ implies that $\overline{N_r(p)} \cap \overline{N_r(q)} = \varnothing$, $\langle S, \tau \rangle$ is a $T_{5/2}$-space.

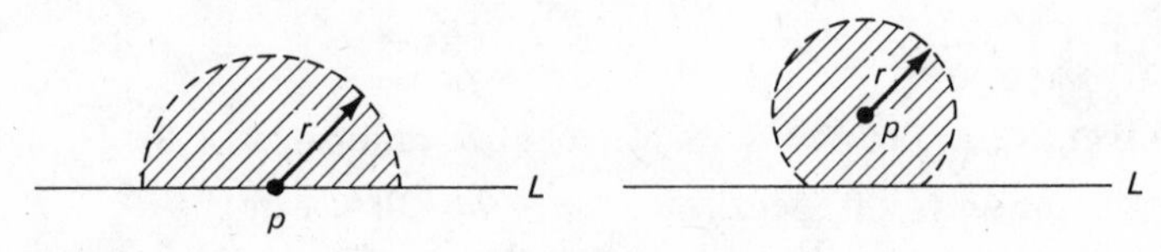

FIGURE 2-6

It is clear from Definitions 2-3 and 2-4 that every $T_{5/2}$-space is a T_2-space and every T_2-space is a T_1-space. The reader can easily check that a metric space is a $T_{5/2}$-space. We show that the property of being a $T_{5/2}$-space is both hereditary and productive, and conclude this section with the very useful result which states that sequential limits in a Hausdorff space are unique.

THEOREM 2-2. *If $\langle S, \tau \rangle$ is a $T_{5/2}$-space, then every subspace of $\langle S, \tau \rangle$ is also a $T_{5/2}$-space.*

Proof. Let $\varnothing \neq A \subset S$ and $p, q \in A$ with $p \neq q$. Since $p, q \in S, \exists U, V \in \tau$ such that $p \in U$, $q \in V$, and $\overline{U} \cap \overline{V} = \varnothing$. Thus $p \in A \cap U$, $q \in A \cap V$, and we have $(\overline{A \cap U} \cap A) \cap (\overline{A \cap V} \cap A) = \varnothing$, since $\overline{U} \cap \overline{V} = \varnothing$. Hence $\langle A, \tau/A \rangle$ is a $T_{5/2}$-space.

THEOREM 2-3. *If $\langle S_\alpha, \tau_\alpha \rangle$ is a $T_{5/2}$-space $\forall \alpha \in \Lambda$, then the product space $\langle S, \tau \rangle = \langle \Pi_\Lambda S_\alpha, \Pi_\Lambda \tau_\alpha \rangle$ is also a $T_{5/2}$-space.*

Proof. Let $p, q \in S$ with $p \neq q$. Thus $p_\alpha \neq q_\alpha$ for some $\alpha \in \Lambda$. Since $\langle S_\alpha, \tau_\alpha \rangle$ is a $T_{5/2}$-space, $\exists U_\alpha, V_\alpha \in \tau_\alpha$ such that $p_\alpha \in U_\alpha$, $q_\alpha \in V_\alpha$, and $\overline{U}_\alpha \cap \overline{V}_\alpha = \varnothing$. Thus $\pi_\alpha^{-1}(U_\alpha), \pi_\alpha^{-1}(V_\alpha) \in \tau$ and $p \in \pi_\alpha^{-1}(U_\alpha)$, $q \in \pi_\alpha^{-1}(V_\alpha)$. Since the projections π_α are all continuous, we have $\overline{\pi_\alpha^{-1}(U_\alpha)} \subset \pi_\alpha^{-1}(\overline{U}_\alpha)$ and $\overline{\pi_\alpha^{-1}(V_\alpha)} \subset \pi_\alpha^{-1}(\overline{V}_\alpha)$. Also $\pi_\alpha^{-1}(\overline{U}_\alpha) \cap \pi_\alpha^{-1}(\overline{V}_\alpha) = \varnothing$, since $\overline{U}_\alpha \cap \overline{V}_\alpha = \varnothing$. Thus $\overline{\pi_\alpha^{-1}(U_\alpha)} \cap \overline{\pi_\alpha^{-1}(V_\alpha)} = \varnothing$, and $\langle S, \tau \rangle$ is a $T_{5/2}$-space.

THEOREM 2-4. *Sequential limits in a Hausdorff space are unique.*

Proof. Let $\langle S, \tau \rangle$ be a Hausdorff space. Suppose that $\{x_n\}_{n \in I^+}$ is a sequence in S with two sequential limits L_1 and L_2. Then $\exists U, V \in \tau$ such that $L_1 \in U$, $L_2 \in V$, and $U \cap V = \varnothing$. Thus $\exists N_1, N_2 \in I^+$ such that $x_n \in U, \forall n \geq N_1$ and $x_n \in V, \forall n \geq N_2$. This implies that $x_n \in U \cap V$, $\forall n \geq \max\{N_1, N_2\}$. Contradiction.

EXERCISES

2-1. Show that the property of being a T_i-space ($i = 0, 1, 2, 5/2$) is a topological property that is both hereditary and productive.

2-2. Show that $\langle S, \tau \rangle$ is a T_0-space iff $x \neq y$ implies that $\overline{\{x\}} \neq \overline{\{y\}}$ $\forall x, y \in S$.

2-3. Show that the discrete topology is the only T_1-topology on a finite set S.

2-4. Show that if $\langle S, \tau \rangle$ is a T_1-space, $x \in S$ and $A \subset S$, then x is a limit point of A iff every open set containing x contains infinitely many distinct points of A.

2-5. Show that every finite subspace of a T_1-space is closed.

2-6. Let $\langle S, \tau \rangle$ be a first countable T_1-space. If $x \in S$ and $A \subset S$, show that x is a limit point of A iff there exists a sequence of distinct points of A converging to x.

2-7. Let $\langle S, \tau \rangle$ be a first countable space in which sequential limits are unique. Show that $\langle S, \tau \rangle$ is Hausdorff.

2-8. Show that $\langle S, \tau \rangle$ is Hausdorff iff the diagonal $\Delta = \{\langle x, x \rangle \mid x \in S\}$ is a closed subspace of the product space $\langle S \times S, \tau \times \tau \rangle$.

2-9. Let $f: \langle S, \tau_1 \rangle \xrightarrow{\text{onto}} \langle T, \tau_2 \rangle$ be a function such that f is 1-1 and f^{-1} is continuous. If $\langle S, \tau_1 \rangle$ is Hausdorff, show that $\langle T, \tau_2 \rangle$ is also Hausdorff.

2-10. Let $f: \langle S, \tau_1 \rangle \xrightarrow{\text{onto}} \langle T, \tau_2 \rangle$ be a closed function and $\langle S, \tau_1 \rangle$ a T_1-space. Show that $\langle T, \tau_2 \rangle$ is a T_1-space also.

2-11. We define $\langle S, \tau \rangle$ to be a "T_D-space" iff $\{x\}'$ is closed $\forall x \in S$. Show that the property of being a T_D-space is a hereditary topological property that implies T_0 and is implied by T_1.

2-2 Regular (T_3) and Completely Regular ($T_{7/2}$)

In this section we consider the axioms relating to the separation of a point from a closed set not containing that point. Since these two axioms, "regularity" and "complete regularity," do not imply $T_{5/2}$, we define the concept of a "T_3-space" (regular T_1-space) and the concept of a "Tychonoff or $T_{7/2}$-space" (completely regular T_1-space). It follows that $T_{7/2}$ implies T_3 implies $T_{5/2}$.

DEFINITION 2-5. $\langle S, \tau\rangle$ is *regular* iff $p \in S$ and $F \subset S - \{p\}$ with F closed implies that $\exists U, V \in \tau$ with $p \in U$, $F \subset V$, and $U \cap V = \varnothing$. $\langle S, \tau\rangle$ is a T_3-space iff $\langle S, \tau\rangle$ is a regular T_1-space. See Figure 2-7.

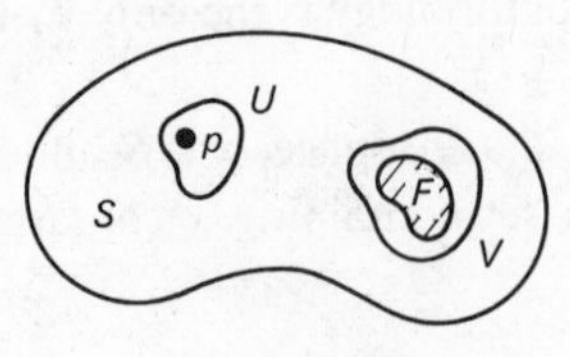

FIGURE 2-7

The space $\langle S, \tau\rangle$ described in Example 2-3 is $T_{5/2}$ but not T_3. For let $p = \langle 0, 0\rangle$ and $F = L - \{p\}$. F is closed, but there do not exist *disjoint* open sets U and V containing p and F, respectively. We describe next a space that is regular (T_3), and then characterize regularity in another way.

EXAMPLE 2-4. Let $S = \{\langle x, y\rangle \in R \times R \mid y \geq 0\}$. If $\langle p, q\rangle \in S$ and $q > 0$, let $N_\varepsilon(p, q) = \{\langle x, y\rangle \in S \mid (x - p)^2 + (y - q)^2 < \varepsilon^2\}$; for each $\langle p, 0\rangle \in S$, let $N_\varepsilon(p, 0) = \{\langle x, y\rangle \in S \mid (x - p)^2 + (y - \varepsilon)^2 < \varepsilon^2\} \cup \{\langle p, 0\rangle\}$. See Figure 2-8. Let $\mathscr{B} = \{N(p, q) \mid \langle p, q\rangle \in S, \varepsilon > 0\}$. The reader may verify that $\mathscr{B}$ is a base for a topology τ on S. Let $L = \{\langle p, 0\rangle \mid p \in R\}$. Notice that τ/L is the discrete topology and $\tau/S - L$ is the Euclidean topology. As in Example 2-3, the only trouble appears to be in trying to separate $p = \langle 0, 0\rangle$ and $F = L - \{p\}$, which is closed. However, the trouble is only superficial. Let $\varepsilon > 0$ and $U = N_\varepsilon(0, 0)$. For each $\langle r, 0\rangle \in F$, $\exists\varepsilon_r > 0$ such that $N_{\varepsilon_r}(r, 0) \cap N_\varepsilon(0, 0) = \varnothing$, where $\varepsilon_r \to 0$ as $r \to 0$. Clearly, $V = \bigcup\{N_{\varepsilon_r}(r, 0) \mid r \in R - \{0\}\} \supset F$ and $U \cap V = \varnothing$. Thus $\langle S, \tau\rangle$ is regular.

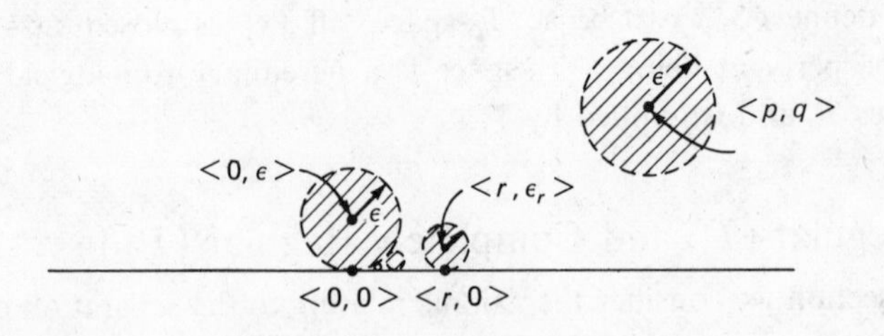

FIGURE 2-8

THEOREM 2-5. *$\langle S, \tau \rangle$ is regular iff $p \in U \in \tau$ implies that $\exists V \in \tau$ such that $p \in V \subset \overline{V} \subset U$.*

Proof. Let $\langle S, \tau \rangle$ be regular and $p \in U \in \tau$. Then $S - U$ is closed and $\exists G_1, G_2 \in \tau$ with $p \in G_1$ and $S - U \subset G_2$ and $G_1 \cap G_2 = \varnothing$. Thus $p \in G_1 \subset S - G_2 \subset U$. Also, $\overline{G}_1 \subset S - G_2$, since $S - G_2$ is closed. Let $V = G_1$.

Now assume that $p \in U \in \tau$ implies $\exists V \in \tau$ such that $p \in V \subset \overline{V} \subset U$. Let $p \in S$ and $F \subset S - \{p\}$, with F closed. Then $p \in S - F \in \tau$. Hence $\exists V \in \tau$ such that $p \in V \subset \overline{V} \subset S - F$. This implies that $F \subset S - \overline{V} \in \tau$ and $(S - \overline{V}) \cap V = \varnothing$. Thus $\langle S, \tau \rangle$ is regular.

Next we discuss the separation of a point p and a closed set $F \subset S - \{p\}$ by means of a continuous real-valued function. This yields the class of completely regular spaces that coincides with the class of uniform and proximity spaces.

DEFINITION 2-6. $\langle S, \tau \rangle$ is *completely regular* iff $p \in S$ and $F \subset S - \{p\}$, with F closed, implies the existence of a continuous function $f: S \to I^1$ with $f(p) = 0$ and $f(F) = \{1\}$. $\langle S, \tau \rangle$ is a *Tychonoff* or *$T_{7/2}$-space* iff $\langle S, \tau \rangle$ is a completely regular T_1-space. See Figure 2-9.

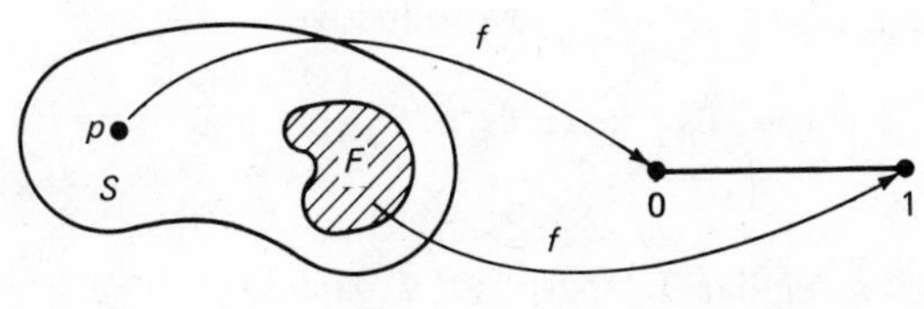

FIGURE 2-9

Most examples of regular spaces are also completely regular. R. F. Arens has constructed an example of a regular space that is not completely regular by gluing together infinitely many copies of a topological space and adjoining two additional points. The details of his construction can be found in Greever, *Theory and Examples of Point-Set Topology*, pp. 77–79. Completely regular spaces have the important property of "uniformizability"; i.e., a completely regular space $\langle S, \tau \rangle$ admits a uniformity $\mathscr{U}$ such that the uniform topology $\tau_{\mathscr{U}}$ is τ.

THEOREM 2-6. *Every completely regular space is uniformizable.*

Proof. Let $\langle S, \tau \rangle$ be completely regular. If $f: S \to I^1$ is continuous and $\varepsilon > 0$, let $B_{f,\varepsilon} = \{\langle x, y \rangle \in S \times S \mid |f(x) - f(y)| < \varepsilon\}$. Observe that each two of these sets $B_{f,\varepsilon}$ have a nonempty intersection. Denote by $\mathscr{B}$ the collection of all intersections of finitely many of these sets $B_{f,\varepsilon}$, and let $\mathscr{U} = \{U \subset S \times S \mid U \supset B \text{ for some } B \in \mathscr{B}\}$. We show that $\mathscr{U}$ is a uniformity. $(U1)$ is satisfied, since $|f(x) - f(x)| = 0 < \varepsilon$ implies that $\langle x, x\rangle \in B_{f,\varepsilon}$, $\forall x \in S$. Clearly, $(U2)$ is satisfied, since $U \in \mathscr{U}$ and $U \subset V$ implies that $\exists B \in \mathscr{B}$ such that $B \subset U \subset V$, and $V \in \mathscr{U}$. If $U, V \in \mathscr{U}$, then $\exists B_1, B_2 \in \mathscr{B}$ such that $U \supset B_1$ and $V \supset B_2$. This implies that $\varnothing \neq B_1 \cap B_2 \in \mathscr{B}$ and $U \cap V \supset B_1 \cap B_2$. Thus $U \cap V \in \mathscr{U}$ and (U3) is satisfied. The triangle inequality implies that $B_{f,\varepsilon/2} \circ B_{f,\varepsilon/2} \subset B_{f,\varepsilon}$. If $U \in \mathscr{U}$, $\exists B = \bigcap\{B_{f_i,\varepsilon_i} \mid i = 1, 2, \ldots, n\} \in \mathscr{B}$ such that $B \subset U$. Moreover, $\bigcap\{B_{f_i,\varepsilon_i/2} \mid i = 1, 2, \ldots, n\} \circ \bigcap\{B_{f_i,\varepsilon_i/2} \mid i = 1, 2, \ldots, n\} \subset \bigcap\{B_{f_i,\varepsilon_i} \mid i = 1, 2, \ldots, n\} = B$. Thus (U4) is satisfied. Finally, (U5) is satisfied, since $B_{f,\varepsilon}^{-1} = B_{f,\varepsilon}$ implies that $B^{-1} = B$, $\forall B \in \mathscr{B}$. Note also that $\tau_{\mathscr{U}} = \tau$.

The converse of Theorem 2-6 is also true: Every uniform space is completely regular. A proof of this result is found in Pervin, *Foundations of General Topology*, pp. 179–81. Thus the class of completely regular spaces coincides with the class of uniform and proximity spaces.

EXERCISES

2-12. Show that the property of being a T_i-space ($i = 3, \frac{7}{2}$) is a topological property that is both hereditary and productive.

2-13. Show that every $T_{7/2}$-space is a T_3-space and every T_3-space is a $T_{5/2}$-space.

2-14. Prove that a regular T_0-space is a T_3-space.

2-3 Normal (T_4) and Completely Normal (T_5)

Next we investigate the axioms relating to the separation of pairs of disjoint closed sets or "separated sets" by disjoint open sets. Since neither of these properties, normality and complete normality, implies regularity, we define the concepts of a "T_4-space" (normal T_1-space) and a "T_5-space" (completely normal T_1-space). It follows that T_5 implies T_4 implies $T_{7/2}$, as the reader can easily verify. We prove two important theorems, due to Urysohn and Tietze, which provide alternative characterizations of normality. Also, we show that complete normality is equivalent to hereditary normality.

DEFINITION 2-7. $\langle S, \tau\rangle$ is *normal* iff for every pair F_1, F_2 of disjoint, closed subsets of S $\exists G_1, G_2 \in \tau$ with $F_1 \subset G_1$, $F_2 \subset G_2$, and $G_1 \cap G_2 = \varnothing$. $\langle S, \tau\rangle$ is a *T_4-space* iff $\langle S, \tau\rangle$ is a normal T_1-space. See Figure 2-10.

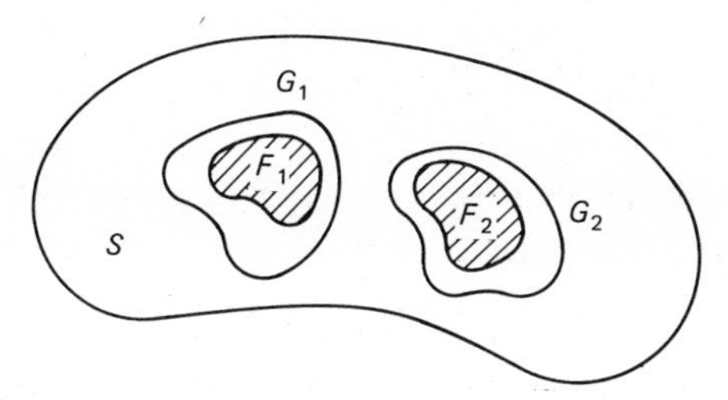

FIGURE 2-10

The space described in Example 2-4 is a $T_{7/2}$-space but is not normal. Let $F_1 = \{\langle r, 0\rangle \mid r \text{ rational}\}$ and $F_2 = \{\langle r, 0\rangle \mid r \text{ irrational}\}$. Clearly, F_1 and F_2 are closed and $F_1 \cap F_2 = \varnothing$. However, there do not exist disjoint open sets G_1 and G_2 containing F_1 and F_2, respectively, since a disk tangent at any irrational (point of F_2) must intersect disks tangent at infinitely many rationals (points of F_1).

THEOREM 2-7 (Urysohn's Lemma). *$\langle S, \tau\rangle$ is normal iff for every pair F_1, F_2 of disjoint, closed subsets of S there exists a mapping $f\colon S \to [a, b]$ such that $f(F_1) = \{a\}$ and $f(F_2) = \{b\}$.*

Proof. If such a mapping $f\colon S \to [a, b]$ exists, then

$$F_1 \subset f^{-1}\left(\left[a, \frac{a+b}{2}\right)\right) = G_1 \in \tau$$

and

$$F_2 \subset f^{-1}\left(\left(\frac{a+b}{2}, b\right]\right) = G_2 \in \tau$$

with $G_1 \cap G_2 = \varnothing$. Thus $\langle S, \tau\rangle$ is normal. Conversely, let $\langle S, \tau\rangle$ be normal and F_1, F_2 be a pair of disjoint closed subsets of S. Let $R^{\#}$ denote the set of rationals and define a collection $\{G_r \mid r \in R^{\#}\}$ of elements $G_r \in \tau$ such that $\overline{G_r} \subset G_s$ if $s > r$ as follows:

(1) If $r < 0$, let $G_r = \varnothing$.

(2) If $r > 1$, let $G_r = S$.

(3) Let $\{r_n\}_{n \in I^+}$ be a labeling of the rationals in $[0, 1]$ with $r_1 = 0$ and $r_2 = 1$. Let $G_1 = S - F_2 \supset F_1$. By Exercise 2-15, the normality of $\langle S, \tau\rangle$ implies that $\exists G_0 \in \tau$ such that $G_0 \supset F_1$ and $\overline{G_0} \subset G_1$. For each rational $0 < r_n < 1$, let r_i be the largest rational in $[0, 1]$ and r_j the smallest rational in $[0, 1]$ such that $i, j < n$ and $r_i < r_n < r_j$. By Exercise 2-15, $\exists G_{r_n} \in \tau$ such that $\overline{G_{r_i}} \subset G_{r_n} \subset \overline{G_{r_n}} \subset G_{r_j}$. Let $g(x) = \inf\{r \in R^{\#} \mid x \in G_r\}$, $\forall x \in S$. Then

$g(F_1) = \{0\}$ and $g(F_2) = \{1\}$. Moreover, g is continuous, since $g^{-1}((p, q)) = \{x \mid g(x) > p\} \cap \{x \mid g(x) < q\} = (\bigcup\{S - \bar{G}_r \mid r > p\}) \cap (\bigcup\{G_r \mid r < q\}) \in \tau$. The function $h: [0, 1] \to [a, b]$, given by $h(x) = (b - a)x + a$, $\forall x \in [0, 1]$, is a homeomorphism. Thus $f = h \circ g: S \to [a, b]$ is continuous with $f(F_1) = \{a\}$ and $f(F_2) = \{b\}$.

THEOREM 2-8 (Tietze's Extension Theorem). *$\langle S, \tau\rangle$ is normal iff each mapping $f: F \to [a, b]$, where $F \subset S$ is closed, has a continuous "extension" $f^*: S \to [a, b]$ [i.e., $f^*(x) = f(x)$, $\forall x \in F$].*

Proof. The function $h: [a, b] \to [-1, 1]$, given by $h(x) = (2x - a - b)/(b - a)$, $\forall x \in [a, b]$, is a homeomorphism. Thus $g = h \circ f: F \to [-1, 1]$ is continuous. We show that g has a continuous extension g^* to S if $\langle S, \tau\rangle$ is normal. Then $h^{-1} \circ g^*$ is the required continuous extension f^* of f to all of S. Let $A_1 = \{x \in F \mid g(x) \geq \frac{1}{3}\}$ and $B_1 = \{x \in F \mid g(x) \leq -\frac{1}{3}\}$. Since g is continuous, A_1 and B_1 are closed and disjoint. By Theorem 2-7 there exists a mapping $g_1: S \to [-\frac{1}{3}, \frac{1}{3}]$ such that $g_1(A_1) = \{\frac{1}{3}\}$ and $g_1(B_1) = \{-\frac{1}{3}\}$. Thus $|g(x) - g_1(x)| \leq \frac{2}{3}$, $\forall x \in F$. Next, let $A_2 = \{x \in F \mid g(x) - g_1(x) \geq \frac{2}{9}\}$ and $B_2 = \{x \in F \mid g(x) - g_1(x) \leq -\frac{2}{9}\}$. Clearly, A_2 and B_2 are closed and disjoint. Hence, by Theorem 2-7, there exists a mapping $g_2: S \to [-\frac{2}{9}, \frac{2}{9}]$ such that $g_2(A_2) = \{\frac{2}{9}\}$ and $g_2(B_2) = \{-\frac{2}{9}\}$. Thus $|g(x) - (g_1(x) + g_2(x))| \leq \frac{4}{9}$, $\forall x \in F$. We continue this process inductively to obtain a sequence of mappings $g_n: S \to [-2^{n-1}/3^n, 2^{n-1}/3^n]$ such that $|g(x) - \sum_{k=1}^{n} g_k(x)| \leq (\frac{2}{3})^n$. Since $\sum_{n=1}^{\infty} 2^{n-1}/3^n = 1$, the Weierstrass M-test implies that

$$\sum_{k=1}^{\infty} g_k$$

converges uniformly to a continuous function $g^*: S \to E^1$ such that $|g^*(x)| \leq 1$, $\forall x \in S$. Since $g(x) - g^*(x) = 0$, $\forall x \in F$, g^* is the required extension of g.

Conversely, suppose that $\langle S, \tau\rangle$ has the "continuous extension property" and let F_1, F_2 be nonempty, disjoint closed subsets of S. We define $f: F_1 \cup F_2 \to [a, b]$ as follows:

$$f(x) = \begin{cases} a & \text{if } x \in F_1 \\ b & \text{if } x \in F_2. \end{cases}$$

Then f is continuous and by hypothesis has a continuous extension f^* to S. Thus, by Urysohn's lemma (Theorem 2-7), $\langle S, \tau\rangle$ is normal.

We discuss next "complete normality," which involves the separation of pairs of "separated sets" by disjoint open sets. We show that completely normal is equivalent to hereditarily normal, and that every metric space is completely normal.

DEFINITION 2-8. Let $\langle S, \tau \rangle$ be a topological space and $A, B \subset S$. A and B are *separated* iff $A \cap \bar{B} = \bar{A} \cap B = \emptyset$.

DEFINITION 2-9. $\langle S, \tau \rangle$ is *completely normal* iff for every pair A, B of separated subsets of S there are disjoint open sets U and V with $A \subset U$ and $B \subset V$. $\langle S, \tau \rangle$ is a *T_5-space* iff $\langle S, \tau \rangle$ is a completely normal T_1-space.

THEOREM 2-9. *$\langle S, \tau \rangle$ is completely normal iff it is hereditarily normal (i.e., each subspace of $\langle S, \tau \rangle$ is normal).*

Proof. Let $\langle S, \tau \rangle$ be completely normal and $\langle A, \tau/A \rangle$ be any subspace of $\langle S, \tau \rangle$. If C_1 and C_2 are disjoint subsets of A that are closed in A, then there are closed subsets F_1 and F_2 of S such that $C_1 = A \cap F_1$ and $C_2 = A \cap F_2$. Thus $\bar{C}_1 \subset F_1$ and $\bar{C}_2 \subset F_2$. This implies that $\bar{C}_1 \cap C_2 = C_1 \cap \bar{C}_2 = \emptyset$, and C_1 and C_2 are separated in S. The complete normality of $\langle S, \tau \rangle$ implies that $\exists G_1, G_2 \in \tau$ such that $C_1 \subset G_1$, $C_2 \subset G_2$, and $G_1 \cap G_2 = \emptyset$. Finally, $C_1 \subset A \cap G_1 \in \tau/A$ and $C_2 \subset A \cap G_2 \in \tau/A$ with $(A \cap G_1) \cap (A \cap G_2) = \emptyset$. This establishes the normality of $\langle A, \tau/A \rangle$. Conversely, let each subspace of $\langle S, \tau \rangle$ be normal, and let C_1 and C_2 be separated in S. If $A = S - \{(\bar{C}_1 - C_1) \cup (\bar{C}_2 - C_2)\}$ has the relative topology τ/A, then C_1 and C_2 are closed in A. The normality of $\langle A, \tau/A \rangle$ implies that $\exists A \cap G_1, A \cap G_2 \in \tau/A$ such that $C_1 \subset A \cap G_1$, $C_2 \subset A \cap G_2$, and $(A \cap G_1) \cap (A \cap G_2) = \emptyset$. This implies that $G_1 \cap G_2 \subset (\bar{C}_1 - C_1) \cup (\bar{C}_2 - C_2)$. Let $U_1 = G_1 \cap (S - \bar{C}_2) \in \tau$ and $U_2 = G_2 \cap (S - \bar{C}_1) \in \tau$. Clearly, $C_1 \subset U_1$ and $C_2 \subset U_2$ with $U_1 \cap U_2 = \emptyset$. This establishes the complete normality of $\langle S, \tau \rangle$.

THEOREM 2-10. *Every metric space is completely normal.*

Proof. Let $\langle S, \rho \rangle$ be a metric space, and let A and B be separated subsets of S. Let $U = \{x \in S \mid \rho(\{x\}, A) < \rho(\{x\}, B)\}$ and $V = \{x \in S \mid \rho(\{x\}, A) > \rho(\{x\}, B)\}$. Since $\bar{A} \cap B = A \cap \bar{B} = \emptyset$, it is clear that $A \subset U$ and $B \subset V$ with $U \cap V = \emptyset$. We must show that $U, V \in \tau_\rho$. Let $x \in U$ and $y \in S\rho(x; \varepsilon)$, where $\varepsilon = \frac{1}{3}(\rho(\{x\}, B) - \rho(\{x\}, A))$.

$$\begin{aligned}
\rho(\{y\}, A) &= \inf \{\rho(y, z) \mid z \in A\} \\
&\le \inf \{\rho(y, x) + \rho(x, z) \mid z \in A\} \\
&\le \varepsilon + \inf \{\rho(x, z) \mid z \in A\} \\
&= \varepsilon + \rho(\{x\}, A) \\
&= \rho(\{x\}, B) - 2\varepsilon \\
&= \inf \{\rho(x, z) \mid z \in B\} - 2\varepsilon \\
&\le \inf \{\rho(x, y) + \rho(y, z) \mid z \in B\} - 2\varepsilon \\
&\le \inf \{\rho(y, z) \mid z \in B\} - \varepsilon \\
&= \rho(\{y\}, B) - \varepsilon \\
&< \rho(\{y\}, B).
\end{aligned}$$

Thus $S_\rho(x; \varepsilon) \subset U$, which implies that $U \in \tau_\rho$. A similar argument shows that $V \in \tau_\rho$ also. This establishes the complete normality of $\langle S, \rho \rangle$.

We conclude this section with several examples. The first example is a completely normal space that is not a T_0-space. The second example (due to Tychonoff) is a normal space that is not completely normal. The third example shows that neither normality nor complete normality are productive. Our final example show that none of the T_i separation properties are continuous invariants.

EXAMPLE 2-5. Let $S = \{a, b, c\}$ and $\tau = \{\varnothing, \{a\}, \{b, c\}, S\}$. $\langle S, \tau \rangle$ is not a T_0-space, since each member of τ that contains b also contains c. $\langle S, \tau \rangle$ is easily seen to be regular, completely regular, and completely normal, however, since each member of τ is also closed.

EXAMPLE 2-6. This example is a modified version of what has been referred to in the literature as the "Tychonoff plank." It is a normal space that is not completely normal. If R is the set of real numbers, then $R - \{1\}$ has a well-ordering $\prec$ by the Well-Ordering Principle. If $x, y \in R - \{1\}$, we let $x \prec^* y$ iff $x \prec y$. Also, we let $x \prec^* 1$ for each $x \in R - \{1\}$. Then $\prec^*$ is a well-ordering for R. Denote by ω the least element of R having denumerably many predecessors in R and by Ω the least element of R having uncountably many predecessors in R. Let $S_1 = \{y \in R \mid y \prec^* \omega\} \cup \{\omega\}$ and $S_2 = \{y \in R \mid y \prec^* \Omega\} \cup \{\Omega\}$, and let $\prec_1^* = \prec^* \mid S_1$ and $\prec_2^* = \prec^* \mid S_2$. Let τ_1 be the topology on S_1 having as a subbase the collection $\mathscr{S}_1 = \{L_p^1 \mid p \in S_1\} \cup \{R_p^1 \mid p \in S_1\}$, where $L_p^1 = \{x \in S_1 \mid x \prec_1^* p\}$ and $R_p^1 = \{x \in S_1 \mid p \prec_1^* x\}$. Similarly, let τ_2 be the topology on S_2 having as a subbase the collection $\mathscr{S}_2 = \{L_p^2 \mid p \in S_2\} \cup \{R_p^2 \mid p \in S_2\}$, where $L_p^2 = \{x \in S_2 \mid x \prec_2^* p\}$ and $R_p^2 = \{x \in S_2 \mid p \prec_2^* x\}$. If $\langle s, \tau \rangle$ denotes the product space $\langle S_2 \times S_1, \tau_2 \times \tau_1 \rangle$, then $\langle S, \tau \rangle$ is normal. However, the subspace $\langle T, \tau \mid T \rangle = \langle S - \langle \Omega, \omega \rangle, \tau \mid S - \langle \Omega, \omega \rangle \rangle$ is not normal. For let $F_1 = (S_2 - \{\Omega\}) \times \{\omega\}$ and $F_2 = \{\Omega\} \times (S_1 - \{\omega\})$. Then F_1 and F_2 are closed, nonempty disjoint subsets of T. If $G \in \tau \mid T$ such that $F_2 \subset G$, then G contains a neighbor-

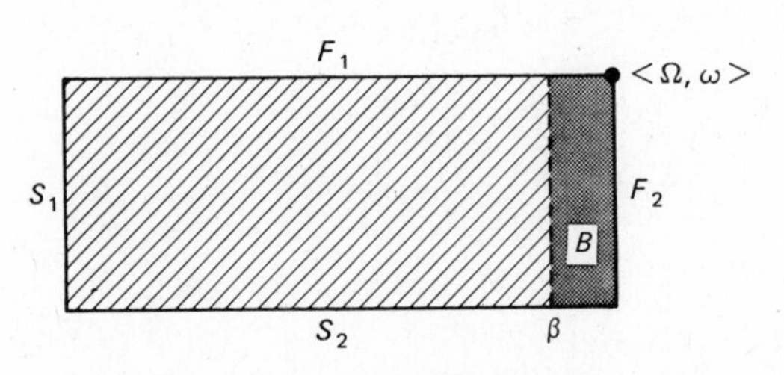

FIGURE 2-11

hood of each of its points. Thus $\exists \beta \in S_2 - \{\Omega\}$ such that $B = \{\langle x, y\rangle \in T \mid \beta \prec_2^* x$ and $y \prec_1^* \omega\} \subset G$. Thus $\bar{G} \cap F_1 \neq \varnothing$ and $\langle T, \tau \mid T\rangle$ is not normal. This implies that $\langle S, \tau\rangle$ is not completely normal.

EXAMPLE 2-7. This is an example of a T_5-space whose product with itself is not normal. In Example 1-2 we described the "lower-limit topology" $\mathscr{L}$ on R. The space $\langle R, \mathscr{L}\rangle$, which is known as the "Sorgenfrey line," is a T_5-space. The verification of this fact is left as an exercise for the reader. Since regularity is productive, $\langle R \times R, \mathscr{L} \times \mathscr{L}\rangle$ is regular (in fact, "Tychonoff"). We show that it is not normal. The sets $A = \{\langle x, -x\rangle \mid x \text{ is rational}\}$ and $B = \{\langle x, -x\rangle \mid x \text{ is irrational}\}$ are closed, disjoint subsets of $R \times R$. Suppose that $U, V \in \mathscr{L} \times \mathscr{L}$ such that $A \subset U$ and $B \subset V$. For each $r \in R$ and $\varepsilon > 0$, let $B(r; \varepsilon) = \{\langle x, y\rangle \in R \times R \mid r \leq x < r + \varepsilon, -r \leq y < -r + \varepsilon\}$. For each $\langle r, -r\rangle \in V$, $\exists \varepsilon_r > 0$ such that $B(r; \varepsilon_r) \subset V$. There exists $\varepsilon > 0$ and $a, b \in R$ with $a < b$ such that in $\langle R, \xi\rangle$ the set $\{x \mid x \text{ is irrational}, \varepsilon_x \geq \varepsilon\}$ is dense in (a, b). Let c be rational such that $a < c < b$ and $B(c; \varepsilon_c) \subset U$ with $\varepsilon_c < \varepsilon$. Then $\exists d$ such that $c < d < b$, $\varepsilon_d \geq \varepsilon$ and $d - c < \varepsilon_c$. Thus $B(c; \varepsilon_c) \cap B(d; \varepsilon_d) \neq \varnothing$, which implies that $U \cap V \neq \varnothing$. See Figure 2-12.

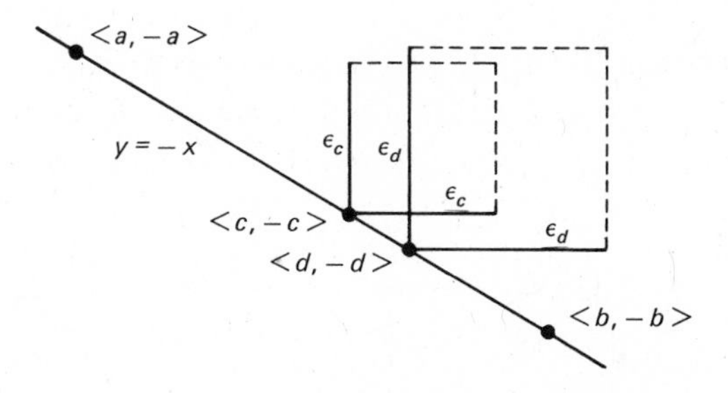

FIGURE 2-12

EXAMPLE 2-8. This is an example of a T_5-space whose continuous image is not even a T_0-space. Let $S = T = \{a, b, c\}$, τ_1 be the discrete topology on S, and $\tau_2 = \{\varnothing, \{a\}, \{b, c\}, T\}$. Since $\langle S, \tau_1\rangle$ is a metrizable space with $d(x, y) = 1$ if $x \neq y$ and 0 otherwise, it is a T_5-space by Theorem 2-10. $\langle T, \tau_2\rangle$ is not a T_0-space, since every open set containing b also contains c, and vice versa. The identity function f given by $f(x) = x$, $\forall x \in S$, is trivially continuous on $\langle S, \tau_1\rangle$, since all functions $g: \langle S, \tau_1\rangle \to \langle T, \tau_2\rangle$ are continuous whenever τ_1 is the discrete topology.

EXERCISES

2-15. Show that $\langle S, \tau\rangle$ is normal iff whenever F is closed a subset of S such that $F \subset U \in \tau$, $\exists V \in \tau$ such that $F \subset V \subset \bar{V} \subset U$.

2-16. Show that $\langle R, \mathscr{L} \rangle$ is a T_5-space.

2-17. If $\langle S, \tau \rangle$ is normal and $f: F \to I^n$, the n-dimensional unit cube $I^1 \times I^1 \times \cdots \times I^1$ (n factors), is continuous with F a closed subset of S, show that f has a continuous extension f^* to all of S.

2-18. Let $f: \langle S, \tau_1 \rangle \xrightarrow{\text{onto}} \langle T, \tau_2 \rangle$ be continuous and closed. Show that if $\langle S, \tau_1 \rangle$ is normal (completely normal), then $\langle T, \tau_2 \rangle$ is normal (completely normal).

★2-4 Collectionwise Normal

We conclude our discussion of the separation properties with a brief mention of "collectionwise normality." This property was introduced and investigated by R. H. Bing in a paper in the *Canadian Journal of Mathematics* in 1951. In this paper he showed that every metric space is collectionwise normal and that every collectionwise normal "Moore space" (defined in Section 3-6) is metrizable. He also gave an example of a normal space that is not collectionwise normal.

DEFINITION 2-10. Let $\langle S, \tau \rangle$ be a topological space. A collection $\mathscr{C}$ of subsets of S is *discrete* iff $\forall x \in S$, $\exists G \in \tau$ such that $x \in G$ and G intersects at most one member of $\mathscr{C}$. $\mathscr{C}$ is *σ-discrete* iff $\mathscr{C} = \bigcup_{n=1}^{\infty} \mathscr{C}_n$ with each $\mathscr{C}_n$ discrete.

DEFINITION 2-11. $\langle S, \tau \rangle$ is *collectionwise normal* iff for each discrete collection $\mathscr{F} = \{F_\alpha \mid \alpha \in \Lambda\}$ of subsets of S there is a class $\mathscr{G} = \{G_\alpha \mid \alpha \in \Lambda\}$ of pairwise disjoint open sets such that $F_\alpha \subset G_\alpha$, $\forall \alpha \in \Lambda$.

Since every pair $\{F_1, F_2\}$ of disjoint, closed sets is a discrete collection, it follows that collectionwise normality implies normality. We defer until Section 3-6 the proof of the collectionwise normality of a metric space, since additional terminology and some preliminary results are necessary.

EXERCISES

2-19. Show that collectionwise normality is a topological property.

2-20. Determine whether or not collectionwise normality is hereditary.

2-21. Let $\langle S, \tau_1 \rangle$ be collectionwise normal and $f: \langle S, \tau_1 \rangle \xrightarrow{\text{onto}} \langle T, \tau_2 \rangle$ continuous. Determine whether or not $\langle T, \tau_2 \rangle$ is necessarily collectionwise normal.

CHAPTER 3
Covering Properties

3-1 Compactness

In this chapter we study the class of covering properties which basically require that an arbitrary "cover" of a space by open sets has specific types of "refinements." First, we introduce the notion of "compactness," which is the strongest of the covering properties we study. In particular, we show that a subspace of $E^1 = \langle R, \xi \rangle$ is compact iff it is closed and bounded (Heine–Borel–Lebesgue theorem). For E^1, compactness is what makes uniform continuity work and assures the existence of maxima and minima. Using the Axiom of Choice (in the form of Zorn's lemma), we prove that compactness is productive (Tychonoff's theorem). J. L. Kelley has shown that Tychonoff's theorem is actually equivalent to the Axiom of Choice.

DEFINITION 3-1. Let $\langle S, \tau \rangle$ be a topological space and $\mathscr{C} = \{G_\alpha \mid \alpha \in \Lambda\} \subset 2^S$. $\mathscr{C}$ is a *cover* of $\langle S, \tau \rangle$ iff $S \subset \bigcup\{G_\alpha \mid \alpha \in \Lambda\}$. A cover $\mathscr{C}$ of $\langle S, \tau \rangle$ is an *open cover* iff $G_\alpha \in \tau$, $\forall \alpha \in \Lambda$. A cover $\mathscr{C}$ of $\langle S, \tau \rangle$ is a *closed cover* iff G_α is closed $\forall \alpha \in \Lambda$.

DEFINITION 3-2. Let $\mathscr{C} = \{G_\alpha \mid \alpha \in \Lambda\}$ be a cover of a topological space $\langle S, \tau \rangle$ and $\mathscr{U} = \{U_\beta \mid \beta \in \Gamma\}$ be another cover of $\langle S, \tau \rangle$. $\mathscr{U}$ is a *subcover* of $\mathscr{C}$ for $\langle S, \tau \rangle$ iff $U_\beta \in \mathscr{C}$, $\forall \beta \in \Gamma$. More generally, $\mathscr{U}$ is a *refinement* of $\mathscr{C}$ iff $\forall \beta \in \Gamma\ \exists \alpha(\beta) \in \Lambda$ such that $U_\beta \subset G_{\alpha(\beta)}$.

DEFINITION 3-3. $\langle S, \tau \rangle$ is *compact* iff each open cover of $\langle S, \tau \rangle$ contains a finite (open) subcover.

Before investigating this property of "compactness" we should note that whether or not $\langle S, \tau \rangle$ is compact depends almost entirely on the topology τ and not upon the set S—the exception being that if S is a finite set, then $\langle S, \tau \rangle$ is compact for all possible topologies τ. Why? Example 3-1 examines the question of compactness in the case of the set of real numbers with various topologies.

EXAMPLE 3-1. $\langle R, \xi \rangle$ is *not* compact, since the open cover $\mathscr{C} = \{(-n, n) \mid n \in I^+\}$ of $\langle R, \xi \rangle$ does not contain a finite subcover. Also, $\langle R, \mathscr{L} \rangle$ is *not* compact, since the open cover $\mathscr{C} = \{[-n, n) \mid n \in I^+\}$ of $\langle R, \mathscr{L} \rangle$ does not contain a finite subcover. Moreover, $\langle R, \text{discrete}\rangle = \langle R, 2^R\rangle$ is *not* compact, since the open cover $\{\{r\} \mid r \in R\}$ of $\langle R, 2^R\rangle$ is irreducible. However, $\langle R, \text{cofinite}\rangle$ is compact, since if G_α is any member of an open cover of $\langle R, \text{cofinite}\rangle$, then $R - G_\alpha$ is finite and thus contained in the union of finitely many other members of the cover. Similarly, any subspace of $\langle R, \text{cofinite}\rangle$ is compact.

If a and b are real numbers with $a < b$, then $[a, b]$ is a compact subspace of $\langle R, \xi\rangle$ as a consequence of the Heine–Borel–Lebesgue theorem, which we shall establish shortly. However, $[a, b)$ is not compact, since

$$\left\{\left[a, b - \frac{b-a}{2n}\right) \,\middle|\, n \in I^+\right\}$$

is an open cover of $[a, b)$, which does not contain a finite subcover. Thus compactness is not hereditary. We do show that closed subspaces of compact spaces are compact and that compact subspaces of a Hausdorff (T_2) space are closed.

THEOREM 3-1. *If $\langle S, \tau\rangle$ is compact and $A \subset S$ is closed, then $\langle A, \tau \mid A\rangle$ is compact.*

Proof. Let $\mathscr{C}_A = \{G_\alpha \cap A \mid \alpha \in \Lambda\}$ be any (relatively) open cover of $\langle A, \tau \mid A\rangle$. Then $\mathscr{C} = \{G_\alpha \mid \alpha \in \Lambda\} \cup \{S - A\}$ is an open cover of $\langle S, \tau\rangle$. Since $\langle S, \tau\rangle$ is compact, $\mathscr{C}$ contains a *finite* subcover $\mathscr{C}' = \{G_{\alpha_1}, G_{\alpha_2}, \ldots, G_{\alpha_n}\}$ of $\langle S, \tau\rangle$. Clearly, $\{G_{\alpha_i} \cap A \mid i = 1, 2, \ldots, n\}$ is a finite subcover of $\langle A, \tau \mid A\rangle$.

THEOREM 3-2. *If $\langle S, \tau\rangle$ is Hausdorff and $\langle A, \tau \mid A\rangle$ is any compact subspace of $\langle S, \tau\rangle$, then A is closed.*

Proof. Let $q \in S - A$. Since $\langle S, \tau\rangle$ is Hausdorff, for every $p \in A$ there are disjoint open sets U_p and V_p with $p \in U_p$ and $q \in V_p$. Then $\mathscr{C} = \{U_p \mid p \in A\}$ is an open cover of A and contains a finite subcover $\mathscr{C}' = \{U_{p_i} \mid i = 1, 2, \ldots, n\}$ of A, since $\langle A, \tau \mid A\rangle$ is compact. Thus $q \in \bigcap\{V_{p_i} \mid i = 1, 2, \ldots, n\} \in \tau$ and $A \cap (\bigcap\{V_{p_i} \mid i = 1, 2, \ldots, n\}) = \varnothing$. Hence A is closed.

EXAMPLE 3-2. $\langle R, \text{cofinite}\rangle$ is hereditarily compact. Thus $\forall r \in R$, $R - \{r\}$ is a nonclosed compact subspace. This shows that the Hausdorff condition in Theorem 3-2 cannot be weakened to T_1.

THEOREM 3-3. *If $f\colon \langle S, \tau_1\rangle \xrightarrow{\text{onto}} \langle T, \tau_2\rangle$ is continuous and $\langle S, \tau_1\rangle$ is compact, then $\langle T, \tau_2\rangle$ is compact.*

Proof. Left to the reader as an exercise.

Before proving the Heine–Borel–Lebesgue theorem, which characterizes the compact subspaces of E^1, we use the DeMorgan formulas to establish an equivalent form of compactness, which is sometimes very useful.

DEFINITION 3-4. A collection $\mathscr{C}$ of sets has the *finite intersection property* iff each nonempty finite subcollection of $\mathscr{C}$ has a nonempty intersection.

THEOREM 3-4. *$\langle S, \tau \rangle$ is compact iff for each collection $\mathscr{F} = \{F_\alpha \mid \alpha \in \Lambda\}$ of closed subsets of S which has the finite intersection property we have that $\bigcap\{F_\alpha \mid \alpha \in \Lambda\} \neq \varnothing$.*

Proof. Let $\langle S, \tau \rangle$ be compact and $\mathscr{F} = \{F_\alpha \mid \alpha \in \Lambda\}$ be any collection of closed subsets of S that has the finite intersection property. Suppose that $\bigcap\{F_\alpha \mid \alpha \in \Lambda\} = \varnothing$. Then $S = S - \bigcap\{F_\alpha \mid \alpha \in \Lambda\} = \bigcup\{S - F_\alpha \mid \alpha \in \Lambda\}$. Thus $\mathscr{C} = \{S - F_\alpha \mid \alpha \in \Lambda\}$ is an open cover of $\langle S, \tau \rangle$ and must contain a finite subcover $\mathscr{C}' = \{S - F_{\alpha_i} \mid i = 1, \ldots, n\}$ for $\langle S, \tau \rangle$. Hence $S = \bigcup\{S - F_{\alpha_i} \mid i = 1, \ldots, n\} = S - \bigcap\{F_{\alpha_i} \mid i = 1, \ldots, n\}$, which implies that $\bigcap\{F_{\alpha_i} \mid i = 1, \ldots, n\} = \varnothing$. This contradicts our assumption that $\mathscr{F}$ has the finite intersection property.

Conversely, suppose that $\bigcap\{F_\alpha \mid \alpha \in \Lambda\} \neq \varnothing$ whenever $\mathscr{F} = \{F_\alpha \mid \alpha \in \Lambda\}$ is a collection of closed subsets of S that has the finite intersection property. Assume that $\langle S, \tau \rangle$ is not compact. Then there exists an open cover $\mathscr{C} = \{G_\alpha \mid \alpha \in \Lambda\}$ for $\langle S, \tau \rangle$, which contains no finite subcover for $\langle S, \tau \rangle$. Thus $S - \bigcup\{G_{\alpha_i} \mid i = 1, \ldots, n\} = \bigcap\{S - G_{\alpha_i} \mid i = 1, \ldots, n\} \neq \varnothing$ for every finite subcollection of $\mathscr{C}$; i.e., $\mathscr{F} = \{S - G_{\alpha_i} \mid i = 1, \ldots, n\}$ is a collection of closed subsets of S that has the finite intersection property. Thus $\bigcap\{S - G_\alpha \mid \alpha \in \Lambda\} = S - \bigcup\{G_\alpha \mid \alpha \in \Lambda\} \neq \varnothing$, which contradicts the fact that $\mathscr{C}$ is a cover of $\langle S, \tau \rangle$.

Recall that a set is bounded iff it has both an upper bound and a lower bound. Thus a subset A of E^1 is bounded iff there exist real numbers a and b with $a < b$ such that $A \subset (a, b)$. We next establish that $[a, b]$ is a compact subspace of E^1 and use this to prove the Heine–Borel–Lebesgue theorem.

LEMMA. *If $a, b \in R$ with $a < b$, then $[a, b]$ is a compact subspace of E^1.*

Proof. Let $\mathscr{C}$ be any open cover of $[a, b]$ and $A = \{x \in [a, b] \mid \mathscr{C}$ contains a finite subcover for $[a, x]\}$. Clearly, $a \in A$ and b is an upper bound for A. Thus A has a least upper bound c by the completeness property of E^1 and $a < c \leq b$. Since $\mathscr{C}$ covers $[a, b]$, there exists a $G \in \mathscr{C}$ such that $c \in G$. If $d \in (a, c) \cap G$, then $\mathscr{C}$ contains a finite subcover $\mathscr{C}'$ for $[a, d]$, and thus $\mathscr{C}' \cup \{G\}$ covers $[a, c]$. If $c < b$, then there would exist an $e \in G$ such that

$(c, e] \subset G$ and $\mathscr{C}' \cup \{G\}$ covers $[a, e]$. This implies that $e \in A$ and $e > c$, which contradicts the definition of c as the least upper bound of A. Thus $c = b$ and $[a, b]$ is compact.

THEOREM 3-5 (Heine–Borel–Lebesgue). *$\langle A, \xi \mid A\rangle$ is a compact subspace of $E^1 = \langle R, \xi\rangle$ iff A is closed and bounded.*

Proof. Let $\langle A, \xi \mid A\rangle$ be any compact subspace of E^1. By Theorem 3-2, A is closed, since E^1 is Hausdorff. If A were not bounded, then $\forall n \in I^+$ $A \not\subset (-n, n)$, but $A \subset \bigcup\{(-n, n) \mid n \in I^+\}$. Thus the open cover $\{(-n, n) \mid n \in I^+\}$ for $\langle A, \xi \mid A\rangle$ would not contain any finite subcover for $\langle A, \xi \mid A\rangle$, which would contradict the compactness of $\langle A, \xi \mid A\rangle$. Conversely, suppose that A is closed and bounded. Then $A \subset (-n, n)$ for some $n \in I^+$ and is a closed subspace of $[-n, n]$, which is compact by our lemma. Thus $\langle A, \xi \mid A\rangle$ is compact by Theorem 3-1.

We should note that the above characterization of the compact subspaces of E^1 is peculiar to the interval (Euclidean) topology on R. The reader may easily verify that $[a, b)$ is not a compact subspace of $\langle R, \mathscr{L}\rangle$ but is closed and bounded. In view of Theorem 3-1, $[a, b]$ is not a compact subspace of $\langle R, \mathscr{L}\rangle$ either.

We are now ready to establish the main result of this section, the Tychonoff product theorem. Our proof of it is that of J. L. Kelley and is obtained by using the following result of J. S. Alexander.

THEOREM 3-6 (Alexander). *$\langle S, \tau\rangle$ is compact iff τ has a subbase $\mathscr{S}$ such that if $\mathscr{C} \subset \mathscr{S}$ covers $\langle S, \tau\rangle$, then $\mathscr{C}$ contains a finite subcover for $\langle S, \tau\rangle$.*

Proof. If $\langle S, \tau\rangle$ is compact, let $\mathscr{S} = \tau$ and the result is trivial. Conversely, let $\mathscr{S}$ be a subbase for τ satisfying the hypothesis of the theorem and suppose that $\mathscr{A} \subset \tau$ has the property that no finite subcollection of $\mathscr{A}$ covers $\langle S, \tau\rangle$. We show that $\langle S, \tau\rangle$ is compact by showing that $\mathscr{A}$ does not cover $\langle S, \tau\rangle$. Let $\mathscr{C} = \{\mathscr{A}' \subset \tau \mid \mathscr{A} \subset \mathscr{A}'$ and no finite subcollection of $\mathscr{A}'$ covers $\langle S, \tau\rangle\}$. Clearly, $\mathscr{A} \in \mathscr{C}$ and $\mathscr{C}$ is partially ordered by inclusion. Let $\mathscr{D} = \{\mathscr{D}_\alpha \mid \alpha \in \Lambda\}$ be any chain in $\langle \mathscr{C}, \subset\rangle$, and let $\mathscr{U} = \{D \mid D \in \mathscr{D}_\alpha$ for some $\alpha \in \Lambda\}$. Then $\mathscr{U}$ is an upper bound for $\mathscr{D}$ and $\mathscr{U} \in \mathscr{C}$. Hence $\mathscr{C}$ has a maximal element $\mathscr{M}$ by Zorn's lemma. Since $\mathscr{A} \subset \mathscr{M}$, we need only show that $\mathscr{M}$ does not cover $\langle S, \tau\rangle$ and the theorem is proved. Let $x \in M \in \mathscr{M}$. Since $\mathscr{S}$ is a subbase for τ, there exists $\{S_1, S_2, \ldots, S_n\} \subset \mathscr{S} \subset \tau$ such that $x \in \bigcap\{S_i \mid i = 1, \ldots, n\} \subset M$. Suppose that $S_i \notin \mathscr{M}$, $\forall i = 1, \ldots, n$. Since $\mathscr{M}$ is a maximal element of $\mathscr{C}$, some finite subcollection $\mathscr{M}_i \cup \{S_i\}$ of $\mathscr{M} \cup \{S_i\}$ must cover $\langle S, \tau\rangle$ for each $i = 1, \ldots, n$. Since $\bigcap\{S_i \mid i = 1, \ldots, n\} \subset M$,

$\{M\} \cup \mathscr{M}_1 \cup \cdots \cup \mathscr{M}_n$ covers $\langle S, \tau \rangle$. Thus $M \notin \mathscr{M}$, since no finite subcollection of $\mathscr{M}$ covers $\langle S, \tau \rangle$. Contradiction. Hence $x \in S_i \in \mathscr{M}$ for *some* $i \in \{1, \ldots, n\}$ and $\bigcup\{M \mid M \in \mathscr{M}\} \subset \bigcup\{S_\alpha \cap M \mid S_\alpha \in \mathscr{S}, \mathscr{M} \in M\}$. If $\mathscr{M}$ were a cover for $\langle S, \tau \rangle$, then $\{S_\alpha \cap M \mid S_\alpha \in \mathscr{S}, M \in \mathscr{M}\}$ would also cover $\langle S, \tau \rangle$. Hence some finite subcollection of $\{S_\alpha \cap M \mid S_\alpha \in \mathscr{S}, M \in \mathscr{M}\}$ would also cover $\langle S, \tau \rangle$ by hypothesis. This would imply that some finite subcollection of $\mathscr{M}$ would cover $\langle S, \tau \rangle$, which is a contradiction.

THEOREM 3-7 (Tychonoff). *If* $\langle S_\alpha, \tau_\alpha \rangle$ *is compact* $\forall \alpha \in \Lambda$, *then* $\langle \prod_\Lambda S_\alpha, \prod_\Lambda \tau_\alpha \rangle$ *is compact also.*

Proof. By Definition 1-21, $\mathscr{S} = \{\pi_\alpha^{-1}(G_\alpha) \mid G_\alpha \in \tau_\alpha, \alpha \in \Lambda\}$ is a subbase for $\prod_\Lambda \tau_\alpha$. Suppose that $\mathscr{A} \subset \mathscr{S}$ and no finite subcollection of $\mathscr{A}$ covers $\langle \prod_\Lambda S_\alpha, \prod_\Lambda \tau_\alpha \rangle$. $\forall \alpha \in \Lambda$, let $\mathscr{C}_\alpha = \{G_\alpha \in \tau_\alpha \mid \pi_\alpha^{-1}(G_\alpha) \in \mathscr{A}\}$. No finite subcollection of $\mathscr{C}_\alpha$ covers $\langle S_\alpha \tau_\alpha \rangle$, since otherwise some finite subcollection of $\mathscr{A}$ would cover $\langle \prod_\Lambda S_\alpha, \prod_\Lambda \tau_\alpha \rangle$. Moreover, since $\langle S_\alpha, \tau_\alpha \rangle$ is compact, $\mathscr{C}_\alpha$ is not a cover for $\langle S_\alpha, \tau_\alpha \rangle$ $\forall \alpha \in \Lambda$. Hence $\exists x_\alpha \in S_\alpha - \bigcup\{G_\alpha \mid G_\alpha \in \mathscr{C}_\alpha\}$, $\forall \alpha \in \Lambda$. Thus the function $x : \Lambda \to \bigcup\{S_\alpha \mid \alpha \in \Lambda\}$ given by $x(\alpha) = x_\alpha$ is in $\prod_\Lambda S_\alpha$ but is *not* in $\bigcup\{A \mid A \in \mathscr{A}\}$. By Theorem 3-6, $\langle \prod_\Lambda S_\alpha, \prod_\Lambda \tau_\alpha \rangle$ is compact since $\mathscr{A}$ does not cover it.

EXERCISES

3-1. Prove Theorem 3-3.

3-2. Show that the union of finitely many compact subspaces of a topological space is compact.

3-3. Let $\langle S, \tau_1 \rangle$ be compact and $\langle T, \tau_2 \rangle$ Hausdorff. Show that $f: \langle S, \tau_1 \rangle \xrightarrow{\text{onto}} \langle T, \tau_2 \rangle$ is a homeomorphism iff f is 1-1 and continuous.

3-4. If $\langle S, \tau \rangle$ is a Hausdorff space and P and Q are disjoint compact subspaces of $\langle S, \tau \rangle$, prove that there exist disjoint open sets U and V such that $P \subset U$ and $Q \subset V$.

3-5. Prove that if $\langle S, \tau \rangle$ is a compact Hausdorff space, then $\langle S, \tau \rangle$ is normal (hence T_4).

3-6. $\langle S, \tau \rangle$ is *pseudocompact* iff each continuous function $f: \langle S, \tau \rangle \to E^1$ is bounded (i.e., $\exists M > 0$ such that $|f(x)| \leq M$, $\forall x \in S$). Show that compact implies pseudocompact and give a counterexample to show that the converse is false.

3-7. $\langle S, \tau \rangle$ is *locally compact* iff $\forall p \in S\ \exists G_p \in \tau$ and a compact subspace K_p of $\langle S, \tau \rangle$ such that $p \in G_p \subset K_p$. Establish the validity of the following statements about local compactness.

(a) A closed subspace of a locally compact space is locally compact.

(b) Compactness implies local compactness. (Give an example of a locally compact space that is not compact.)

(c) If each point $p \in S$ is contained in an open set G_p such that $\overline{G_p}$ is compact, then $\langle S, \tau \rangle$ is locally compact. Is the converse true?

(d) If $f: \langle S, \tau_1 \rangle \xrightarrow{\text{onto}} \langle T, \tau_2 \rangle$ is continuous and open, and $\langle S, \tau_1 \rangle$ is locally compact, then $\langle T, \tau_2 \rangle$ is locally compact.

(e) The product of finitely many locally compact spaces is locally compact.

3-8. Let $\mathscr{F}(S, T)$ denote the collection of all functions from $\langle S, \tau_1 \rangle$ into $\langle T, \tau_2 \rangle$. For each compact $K \subset S$ and $\forall G \in \tau_2$, let $N_{K,G} = \{f \in \mathscr{F}(S, T) \mid f(K) \subset G\}$. Show that $\mathscr{S} = \{N_{K,G} \mid K^{\text{compact}} \subset S, G \in \tau_2\}$ is a subbase for a topology on $\mathscr{F}(S, T)$ that we call the *compact-open topology* or the *topology of compact convergence*. This topology will be utilized in our study of homotopy theory in Chapter 7.

3-2 Lindelöf Spaces

In this section we investigate a weaker covering property than compactness known as the "Lindelöf property." The Lindelöf property is also implied by second countability and is equivalent to second countability and separability in a metric space.

DEFINITION 3-5. $\langle S, \tau \rangle$ is *Lindelöf* iff every open cover for $\langle S, \tau \rangle$ contains a countable (open) subcover for $\langle S, \tau \rangle$.

Clearly, compactness implies the Lindelöf property. We examine now our basic topologies on R to see which of them have the Lindelöf property.

EXAMPLE 3-3. $\langle R, \text{cofinite} \rangle$ is compact (hence Lindelöf). $E^1 = \langle R, \xi \rangle$ is hereditarily Lindelöf, since it is second countable. Also, $\langle R, \mathscr{L} \rangle$ is hereditarily Lindelöf even though it is only first countable. $\langle R, \text{discrete} \rangle$ is not Lindelöf, since the open cover of $\langle R, \text{discrete} \rangle$ by the singletons is irreducible. Let $\tau = \{G \subset R \mid \text{either } 0 \notin G \text{ or } R - \{1, 2\} \subset G\}$. Then each open cover for $\langle R, \tau \rangle$ contains a finite subcover for $\langle R, \tau \rangle$. Thus $\langle R, \tau \rangle$ is compact (hence Lindelöf). $\langle R - \{0\}, \tau \mid R - \{0\} \rangle$ is an uncountable subspace of $\langle R, \tau \rangle$, which is not Lindelöf, since the open cover $\{\{r\} \mid r \in R - \{0\}\}$ is irreducible. This shows that the Lindelöf property is not hereditary.

THEOREM 3-8. *If $\langle S, \tau \rangle$ is Lindelöf and $A \subset S$ is closed, then the subspace $\langle A, \tau \mid A \rangle$ is Lindelöf.*

Proof. Left to the reader as an exercise.

THEOREM 3-9. *If $\langle S, \tau_1 \rangle$ is Lindelöf and $f: \langle S, \tau_1 \rangle \xrightarrow{\text{onto}} \langle T, \tau_2 \rangle$ is continuous, then $\langle T, \tau_2 \rangle$ is Lindelöf.*

Proof. Left to the reader as an exercise.

We next give an example of a Lindelöf space whose product with itself is not Lindelöf. Thus the Lindelöf property is not even finitely productive.

EXAMPLE 3-4. As remarked earlier, $\langle R, \mathscr{L} \rangle$ is first countable, separable, and Lindelöf but is not second countable. $\langle R \times R, \mathscr{L} \times \mathscr{L} \rangle$ is first countable and separable but not Lindelöf. Let $Y = \{\langle x, -x \rangle \mid x \in R\}$, and $\forall p \in R \times R$, let N_p be a left-lower box with corner p that intersects Y iff $p \in Y$. Then $\mathscr{C} = \{N_p \mid p \in R \times R\}$ is an open cover for $\langle R \times R, \mathscr{L} \times \mathscr{L} \rangle$ with the property that any subcover for $\langle R \times R, \mathscr{L} \times \mathscr{L} \rangle$ must contain uncountably many members in order to cover up the uncountable subspace Y.

Now we investigate the relationship of the Lindelöf property to separability and first and second countability.

THEOREM 3-10. *If $\langle S, \tau \rangle$ is second countable, then $\langle S, \tau \rangle$ is Lindelöf.*

Proof. Let $\mathscr{B} = \{B_n \mid n \in I^+\}$ be a countable base for τ, and let $\mathscr{C} = \{G_\alpha \mid \alpha \in \Lambda\}$ be any open cover for $\langle S, \tau \rangle$. For each $p \in S$, $\exists G_\alpha \in \mathscr{C}$ such that $p \in G_\alpha$. Hence $\exists B_n \in \mathscr{B}$ such that $p \in B_n \subset G_\alpha$. Select such a G_α for each $p \in S$. The collection of such G_α's is necessarily a countable subcover for $\langle S, \tau \rangle$, since $\mathscr{B}$ is a countable cover for $\langle S, \tau \rangle$.

EXAMPLE 3-5. $\langle R, \text{cofinite} \rangle$ is separable and hereditarily compact (hence hereditarily Lindelöf) but is not first countable. For suppose that $\{G_n \mid n \in I^+\}$ were a countable local base at $x \in R$. Since $R - G_n$ is finite $\forall n \in I^+$, $\bigcup\{R - G_n \mid n \in I^+\}$ is countable. Thus, if $x \in G \in \tau$ $\exists G_n$ such that $x \in G_n \subset G$, which implies that $R - G \subset R - G_n$; i.e., each finite subset of R that does *not* contain x is a subset of $\bigcup\{R - G_n \mid n \in I^+\}$, which implies that $\bigcup\{R - G_n \mid n \in I^+\}$ must be uncountable. Contradiction.

EXAMPLE 3-6. The space $\langle R, \tau \rangle$ of Example 3-3 is compact (hence Lindelöf) and is first countable but not separable. If $x \neq 0$, let $\mathscr{B}_x = \{\{x\}\}$ and if $x = 0$, let $\mathscr{B}_x = \{R - \{1, 2\}\}$. Then $\mathscr{B}_x$ is a countable local base at x $\forall x \in R$. $\langle R, \tau \rangle$ is not separable, since $R - \{0\}$ has the discrete (subspace) topology and is uncountable.

Our final result in this section asserts that the Lindelöf property is a connective link between regularity and normality.

THEOREM 3-11. *If* $\langle S, \tau \rangle$ *is a regular Lindelöf space, then* $\langle S, \tau \rangle$ *is normal.*

Proof. Let F_1 and F_2 be any two disjoint closed subsets of S. Then $\langle F_1, \tau \mid F_1 \rangle$ and $\langle F_2, \tau \mid F_2 \rangle$ are Lindelöf by Theorem 3-8. By regularity, $\forall p \in F_1$, $\exists G_1(p)$ such that $p \in G_1(p) \subset \overline{G_1(p)} \subset S - F_2$. The collection $\{G_1(p) \mid p \in F_1\}$ is an open cover for F_1 and must contain some countable subcover $\{G_1^n \mid n \in I^+\}$. Similarly, $\forall q \in F_2$, $\exists G_2(q) \in \tau$ such that $q \in G_2(q) \subset \overline{G_2(q)} \subset S - F_1$. The collection $\{G_2(q) \mid q \in F_2\}$ is an open cover for F_2 and must contain a countable subcover $\{G_2^n \mid n \in I^+\}$. Let $G_1 = \bigcup\{G_1^n - \bigcup\{\overline{G_2^i} \mid i = 1, \ldots, n\} \mid n \in I^+\}$ and $G_2 = \bigcup\{G_2^n - \bigcup\{\overline{G_1^i} \mid i = 1, \ldots, n\} \mid n \in I^+\}$. It is easily seen that G_1 and G_2 are open and disjoint. The reader may verify that $F_1 \subset G_1$ and $F_2 \subset G_2$, as required for normality.

EXERCISES

3-9. Prove Theorem 3-8.

3-10. Prove Theorem 3-9.

3-11. Show that a metric space $\langle M, d \rangle$ is second countable iff it is Lindelöf.

3-3 Countable Compactness

The covering property that we investigate next, countable compactness, is weaker than compactness because it only requires that countable open covers for a space $\langle S, \tau \rangle$ contain finite subcovers for $\langle S, \tau \rangle$. However, countable compactness has many of the same features topologically as does compactness. Indeed, in the context of a metric space or even a Lindelöf space, countable compactness is equivalent to compactness.

DEFINITION 3-6. $\langle S, \tau \rangle$ is *countably compact* iff each countable open cover for $\langle S, \tau \rangle$ contains a finite subcover for $\langle S, \tau \rangle$.

In the terminology of the finite intersection property which was introduced in Definition 3-4, our definition of countable compactness has the following equivalent form. The proof is analogous to that of Theorem 3-4 and is left to the reader as an exercise.

THEOREM 3-12. *$\langle S, \tau \rangle$ is countably compact iff for each countable collection $\mathscr{F} = \{F_n \mid n \in I^+\}$ of closed subsets of S that has the finite intersection property, we have that $\bigcap\{F_n \mid n \in I^+\} \neq \varnothing$.*

From its definition it is clear that compactness implies countable compactness. We now construct an example of a countably compact space which is not compact. To do this we need the following definition of the order topology for a totally ordered set $\langle S, < \rangle$.

DEFINITION 3-7. Let $\langle S, < \rangle$ be a totally ordered set. For each $p \in S$, let $L_p = \{x \in S \mid x < p\}$ and $R_p = \{x \in S \mid p < x\}$. Let $\mathscr{S} = \{L_p \mid p \in S\} \cup \{R_p \mid p \in S\}$. Then $\mathscr{S}$ is a subbase for the *order topology* on S.

EXAMPLE 3-7. Let $\mathscr{O}$ be the set of all ordinals that precede Ω, as in Example 0-9. Let $\mathscr{W}$ denote the order topology for $\mathscr{O}$. Then $\langle \mathscr{O}, \mathscr{W} \rangle$ is countably compact but not Lindelöf (hence not compact). We shall not prove the countable compactness of $\langle \mathscr{O}, \mathscr{W} \rangle$ here, since we shall show later in this chapter that $\langle \mathscr{O}, \mathscr{W} \rangle$ has the stronger covering property of sequential compactness.

EXAMPLE 3-8. $\langle R, \xi \rangle$ and $\langle R, \mathscr{L} \rangle$ are not countably compact, since the respective countable open covers $\{(-n, n) \mid n \in I^+\}$ and $\{[-n, n) \mid n \in I^+\}$ do not contain finite subcovers. Also, $[a, b]$ is a compact subspace of $\langle R, \xi \rangle$ and thus countably compact. However, (a, b) is not countably compact, since the countable open cover

$$\left\{\left(a, b - \frac{b-a}{2n}\right) \mid n \in I^+\right\}$$

does not contain a finite subcover for (a, b).

Although countable compactness is not hereditary, closed subspaces of countably compact spaces are countably compact. Also, countable compactness is a continuous invariant. The proofs of these two results are left as exercises for the reader.

THEOREM 3-13. *If $\langle S, \tau \rangle$ is countably compact and $A \subset S$ is closed, then the subspace $\langle A, \tau \mid A \rangle$ is countably compact.*

THEOREM 3-14. *If $\langle S, \tau_1 \rangle$ is countably compact and $f: \langle S, \tau_1 \rangle \xrightarrow{\text{onto}} \langle T, \tau_2 \rangle$ is continuous, then $\langle T, \tau_2 \rangle$ is countably compact.*

Using the alternative characterization of countable compactness given in Theorem 3-12, we obtain a simple proof of the famous Cantor product theorem concerning the intersection of a sequence of nested closed subspaces of a countably compact space.

THEOREM 3-15. *Let* $\langle S, \tau \rangle$ *be a space and* $\langle C_1, \tau \mid C_1 \rangle$ *a countably compact subspace of* $\langle S, \tau \rangle$. *If* $C_1 \supset C_2 \supset \cdots \supset C_n \supset \cdots$ *is a sequence of nonempty closed subsets of* S, *then* $\bigcap\{C_n \mid n \in I^+\} \neq \emptyset$.

Proof. Since $\{C_n \mid n \in I^+\}$ is a "nested sequence" of nonempty closed subsets of $\langle C_1, \tau \mid C_1 \rangle$, it has the finite intersection property. Thus $\bigcap\{C_n \mid n \in I^+\} \neq \emptyset$ by Theorem 3-12, since $\langle C_1, \tau \mid C_1 \rangle$ is countably compact.

Another equivalent form of countable compactness involves the concept of a "cluster point" of a sequence that we discuss below.

DEFINITION 3-8. $x \in S$ is a *cluster point* of a sequence $\{x_n\}_{n \in I^+}$ in $\langle S, \tau \rangle$ iff whenever $x \in G \in \tau$, $\forall N \in I^+$, we have $G \cap \{x_n \mid n \geq N\} \neq \emptyset$.

THEOREM 3-16. $\langle S, \tau \rangle$ *is countably compact iff each sequence in* $\langle S, \tau \rangle$ *has a cluster point in* S.

Proof. Suppose that $\{x_n\}_{n \in I^+}$ is a sequence in $\langle S, \tau \rangle$ having no cluster point in S. Then $\forall x \in S$ $\exists G_x \in \tau$ and $N \in I^+$ such that $x \in G_x$ and $G_x \cap \{x_{N+1}, x_{N+2}, \ldots\} = \emptyset$. For each $n \in I^+$, let $U_n = \bigcup\{G_x \mid G_x \cap \{x_{n+1}, x_{n+2}, \ldots\} = \emptyset\}$. Then $\{U_n \mid n \in I^+\}$ is a countable open cover for $\langle S, \tau \rangle$, which does not contain any finite subcover for $\langle S, \tau \rangle$. Thus $\langle S, \tau \rangle$ is not countably compact. Conversely, suppose that $\langle S, \tau \rangle$ is not countably compact. Then there exists a countable open cover $\{U_n \mid n \in I^+\}$ for $\langle S, \tau \rangle$, which contains no finite subcover for $\langle S, \tau \rangle$. Let $V_1 = U_1$ and $x_1 \in V_1$. If $n > 1$, let V_n denote the first U_i *not* contained in $\bigcup\{V_i \mid i = 1, \ldots, n-1\}$ and let $x_n \in V_n - \bigcup\{V_i \mid i = 1, \ldots, n-1\}$. Thus, if $x \in S$ $\exists N \in I^+$ such that $x \in V_N$ and $V_N \cap \{x_{N+1}, x_{N+2}, \ldots\} = \emptyset$; i.e., x is not a cluster point of S.

Novák and Terasaka have given examples to show that the product of two countably compact spaces may not be countably compact. Thus countable compactness is not even finitely productive, in contrast to compactness, which is productive. We conclude this section with the statement of another feature of a countably compact space that is of special interest to an analyst. Its proof is left to the reader as an exercise.

THEOREM 3-17. *If $\langle S, \tau \rangle$ is countably compact and $f: \langle S, \tau \rangle \to E^1$ is continuous, then f assumes a maximum (also a minimum) value on S.*

EXERCISES

3-12. Show that $\langle S, \tau \rangle$ is compact iff it is countably compact and Lindelöf.

3-13. Prove Theorem 3-12.

3-14. Prove Theorem 3-13.

3-15. Prove Theorem 3-14.

3-16. Prove Theorem 3-17.

3-17. Show that if $\langle S, \tau_1 \rangle$ is compact and $\langle T, \tau_2 \rangle$ is countably compact, then $\langle S \times T, \tau_1 \times \tau_2 \rangle$ is countably compact.

3-4 Sequential Compactness

Next we investigate a property that is stronger than countable compactness but is equivalent to it for the class of first countable spaces. However, this property neither implies, nor is implied by, compactness. The property is called "sequential compactness," because it is defined in terms of certain requirements of sequences in a space rather than in terms of open covers.

DEFINITION 3-9. $\langle S, \tau \rangle$ is *sequentially compact* iff each sequence in $\langle S, \tau \rangle$ has a subsequence that converges to some point of S.

In view of Theorem 3-16, sequential compactness implies countable compactness because the limit of a convergent subsequence is obviously a cluster point of the original sequence. We describe now a space that is compact (hence countably compact and Lindelöf) but not sequentially compact.

EXAMPLE 3-9. Let $I^1 = \langle [0, 1], \xi \mid [0, 1] \rangle$ and let $\langle S, \tau \rangle$ be the product of *uncountably* many copies of I^1. Denote by T the set of all subsequences of the sequence $1, 2, \ldots, n, \ldots$. Then there exists a bijection $f: R \to T$. Let $r \in R$ and $f(r)$ be the subsequence $n_1^r, n_2^r, \ldots$. If $n = n_i^r$ and i is odd, let $x_n(r) = 0$; otherwise, let $x_n(r) = 1$. Then $x_n : R \to I^1$ is a function and $\{x_n\}_{n \in I^+}$ is a sequence in $\langle S, \tau \rangle$. Suppose that $\{x_{n_i}\}_{i \in I^+}$ is a convergent subsequence. If $r = f^{-1}(n_1, n_2, \ldots, n_i, \ldots)$, then the sequence $\{x_{n_i}(r)\}_{i \in I^+}$

converges to some point in I^1; i.e., the sequence 0, 1, 0, 1, . . . converges in I^1. Contradiction because I^1 is Hausdorff. Thus $\{x_n\}_{n \in I^+}$ has no convergent subsequences and $\langle S, \tau \rangle$ cannot be sequentially compact. However, $\langle S, \tau \rangle$ is compact by the Tychonoff product theorem, since I^1 is compact.

THEOREM 3-18. *If $\langle S, \tau \rangle$ is countably compact and first countable, then $\langle S, \tau \rangle$ is sequentially compact.*

Proof. The proof is left to the reader as an exercise.

As was the case with compactness, countable compactness, and the Lindelöf property, sequential compactness is not hereditary. However, closed subspaces of sequentially compact spaces are sequentially compact. Also, sequential compactness is a continuous invariant. The proofs of these results are also left to the reader as exercises.

THEOREM 3-19. *If $\langle S, \tau \rangle$ is sequentially compact and $A \subset S$ is closed, then the subspace $\langle A, \tau \mid A \rangle$ is sequentially compact.*

THEOREM 3-20. *If $\langle S, \tau_1 \rangle$ is sequentially compact and $f: \langle S, \tau_1 \rangle \xrightarrow{onto} \langle T, \tau_2 \rangle$ is continuous, then $\langle T, \tau_2 \rangle$ is sequentially compact.*

Now we discuss two additional examples that will help us to understand how sequential compactness differs from countable compactness and compactness.

EXAMPLE 3-10. Let $\langle \mathcal{O}, \mathcal{W} \rangle$ be the space of countable ordinals with the order topology and $\langle S, \tau \rangle$ the space described in Example 3-9. Then $\langle \mathcal{O}, \mathcal{W} \rangle$ is countably compact (indeed, sequentially compact) and Hausdorff. $\langle S, \tau \rangle$ is compact and Hausdorff. Hence $\langle \mathcal{O} \times S, \mathcal{W} \times \tau \rangle$ is Hausdorff and also countably compact by Exercise 3-17. Since $\pi_1(\mathcal{O} \times S) = \mathcal{O}$ and $\langle \mathcal{O}, \mathcal{W} \rangle$ is not compact (nor Lindelöf), $\langle \mathcal{O} \times S, \mathcal{W} \times \tau \rangle$ is not compact (nor Lindelöf). Also, since $\pi_2(\mathcal{O} \times S) = S$ and $\langle S, \tau \rangle$ is not sequentially compact, $\langle \mathcal{O} \times S, \mathcal{W} \times \tau \rangle$ is not sequentially compact.

EXAMPLE 3-11. Let $\mathcal{O}^* = \mathcal{O} \cup \{\Omega\}$ and $\mathcal{W}^*$ be the order topology on $\mathcal{O}^*$. Then $\langle \mathcal{O}^*, \mathcal{W}^* \rangle$ is a compact and sequentially compact Hausdorff space having $\langle \mathcal{O}, \mathcal{W} \rangle$ as a *nonclosed* sequentially compact subspace. For let $\mathcal{U}$ be any open cover for $\langle \mathcal{O}^*, \mathcal{W}^* \rangle$. Then $\Omega \in G$ for some $G \in \mathcal{U}$. If $p \in G - \{\Omega\}$, then p has only a countable number of predecessors in $\mathcal{O}^*$. This implies that $\mathcal{U}$ contains a countable subcover for $\mathcal{O}^*$, and $\langle \mathcal{O}^*, \mathcal{W}^* \rangle$ is Lindelöf. Also,

$\langle \mathcal{O}^*, \mathcal{W}^* \rangle$ is sequentially compact and Hausdorff since $\langle \mathcal{O}, \mathcal{W} \rangle$ is sequentially compact and Hausdorff. Hence $\langle \mathcal{O}^*, \mathcal{W}^* \rangle$ is compact. Clearly, Ω is a limit point of $\mathcal{O}$ in $\langle \mathcal{O}^*, \mathcal{W}^* \rangle$, which implies that $\langle \mathcal{O}, \mathcal{W} \rangle$ is not closed in $\langle \mathcal{O}^*, \mathcal{W}^* \rangle$.

Finally, we consider sequential compactness in the context of a metric space. We define the concept of a "totally bounded" metric space and show that a sequentially compact metric space is totally bounded (hence separable).

DEFINITION 3-10. Let $\langle M, d \rangle$ be a metric space and $\varepsilon > 0$. An *ε-net* for $\langle M, d \rangle$ is a finite subset F of M such that $\forall x \in M \; \exists y \in F$, with $d(x, y) < \varepsilon$.

DEFINITION 3-11. A metric space $\langle M, d \rangle$ is *totally bounded* or *precompact* iff there exists an ε-net for $\langle M, d \rangle$ for every $\varepsilon > 0$.

THEOREM 3-21. *If $\langle M, d \rangle$ is a sequentially compact metric space, then $\langle M, d \rangle$ is totally bounded.*

Proof. If $\langle M, d \rangle$ is not totally bounded, then there is some fixed $\varepsilon > 0$ such that $\langle M, d \rangle$ has no ε-net. Let $x_1 \in M$. Since $\{x_1\}$ is not an ε-net for $\langle M, d \rangle$, $\exists x_2 \in M$ such that $d(x_1, x_2) \geq \varepsilon$. Since $\{x_1, x_2\}$ is not an ε-net for $\langle M, d \rangle$, $\exists x_3 \in M$ such that $d(x_1, x_3) \geq \varepsilon$ and $d(x_2, x_3) \geq \varepsilon$. If $\{x_1, x_2, \ldots, x_n\} \subset M$ has already been defined such that $d(x_i, x_j) \geq \varepsilon$ whenever $i \neq j$, then $\{x_1, x_2, \ldots, x_n\}$ is not an ε-net for $\langle M, d \rangle$ either. Hence $\exists x_{n+1} \in M$ such that $d(x_i, x_{n+1}) \geq \varepsilon$ for $i = 1, 2, \ldots, n$. Thus we have defined inductively a sequence $\{x_n\}_{n \in I^+}$ in $\langle M, d \rangle$ such that $d(x_i, x_j) \geq \varepsilon$ whenever $i \neq j$. This implies that no subsequence of $\{x_n\}_{n \in I^+}$ can converge and $\langle M, d \rangle$ is not sequentially compact. Contradiction.

COROLLARY. *Every sequential compact metric space is separable.*

Proof. Let $\langle M, d \rangle$ be sequentially compact. By Theorem 3-21, $\langle M, d \rangle$ is totally bounded. Let F_n be a $1/n$-net for $\langle M, d \rangle$ for each $n \in I^+$. Then the set $F = \bigcup \{F_n \mid n \in I^+\}$ is countable and dense in $\langle M, d \rangle$.

EXERCISES

3-18. Prove Theorem 3-18.

3-19. Prove Theorem 3-19.

3-20. Prove Theorem 3-20.

3-21. Show that a metric space $\langle M, d\rangle$ is compact iff $\langle M, d\rangle$ is sequentially compact iff $\langle M, d\rangle$ is countably compact.

3-22. Show that if $\langle S, \tau_1\rangle$ and $\langle T, \tau_2\rangle$ are sequentially compact, then $\langle S \times T, \tau_1 \times \tau_2\rangle$ is sequentially compact.

3-23. A metric space $\langle M, d\rangle$ is bounded iff $\exists b \in R$ such that $d(x, y) \leq b$, $\forall x, y \in M$. Show that if $\langle M, d\rangle$ is any metric space, then there exists a metric ρ on M such that $\langle M, \rho\rangle$ is bounded and the ρ-metric topology on M is identical with the d-metric topology on M.

3-24. Prove that every totally bounded metric space is bounded.

3-5 Bolzano–Weierstrass Property

In this section we discuss a property weaker than countable compactness but which is equivalent to countable compactness for the class of T_1-spaces. This property is known as the "Bolzano–Weierstrass property" and requires that any infinite subspace of such a space have a limit point in the space. For the class of metric spaces, the Bolzano–Weierstrass property is equivalent to compactness, countable compactness, and sequential compactness.

DEFINITION 3-12. $\langle S, \tau\rangle$ has the *Bolzano–Weierstrass property* iff each infinite subset of S has a limit point in S.

We show first that countable compactness implies the Bolzano–Weierstrass property. Then we give an example of a space that has the Bolzano–Weierstrass property but is not countably compact.

THEOREM 3-22. *If $\langle S, \tau\rangle$ is countably compact, then $\langle S, \tau\rangle$ has the Bolzano–Weierstrass property.*

Proof. Suppose that $A \subset S$ is infinite and has no limit point in S. Let $\{x_n\}_{n \in I^+}$ be any sequence of distinct points in A. Let $A_n = \{x_n, x_{n+1}, \ldots\}$ and $B_n = S - A_n$. If $x \in S$, then $\exists G_x \in \tau$ such that $x \in G_x$ and $G_x \cap A$ contains at most one point. Thus $x_n \in G_x$ for at most one value of n. This implies that $G_x \cap A_n = \varnothing$ and $G_x \subset B_n$ for n sufficiently large. For each $n \in I^+$, let U_n denote the interior of B_n. Then $\{U_n \mid n \in I^+\}$ is a countable open cover for $\langle S, \tau\rangle$ and some finite subcollection $\{U_1, \ldots, U_m\}$ must also cover $\langle S, \tau\rangle$, since $\langle S, \tau\rangle$ is countably compact. For $i = 1, \ldots, m + 1$, $x_{m+1} \in A_i$ and thus $x_{m+1} \notin B_i$. Hence $x_{m+1} \notin \bigcup\{U_i \mid i = 1, \ldots, m\} = S$. Contradiction. Hence A must have a limit point in S.

EXAMPLE 3-12. Let I^+ denote the set of positive integers as usual, and let $B_n = \{2n - 1, 2n\}$, $\forall n \in I^+$. If $\mathscr{B} = \{B_n \mid n \in I^+\}$, then $\mathscr{B}$ is a base for a topology τ on I^+. Since $\mathscr{B}$ is countable, $\langle I^+, \tau\rangle$ is second countable (hence separable and Lindelöf). $\langle I^+, \tau\rangle$ is not countably compact, since $\mathscr{B}$ is a denumerable open cover for $\langle I^+, \tau\rangle$, which is irreducible. However, $\langle S, \tau\rangle$ does have the Bolzano–Weierstrass property. For let $A \subset I^+$ be infinite and $n \in A$. If n is odd, then $n + 1 \in A'$. If n is even, then $n - 1 \in A'$.

THEOREM 3-23. *If $\langle S, \tau\rangle$ is a T_1-space having the Bolzano–Weierstrass property, then $\langle S, \tau\rangle$ is countably compact.*

Proof. Left to the reader as an exercise.

$I^1 = \langle [0, 1], \xi \mid [0, 1]\rangle$ has the Bolzano–Weierstrass property, since it is compact. However, the subspace $\langle \{1/n \mid n \in I^+\}, \xi \mid \{1/n| \; n \in I^+\}\rangle$ does not have the Bolzano–Weierstrass property, since the infinite set $\{1/n \mid n \in I^+\}$ does not contain any limit point of itself. Thus the Bolzano–Weierstrass property is not hereditary. However, closed subspaces of a space with the Bolzano–Weierstrass property have the Bolzano–Weierstrass property. Unlike the covering properties discussed in the previous sections, the Bolzano–Weierstrass property is not a continuous invariant, as we show below by counterexample.

THEOREM 3-24. *If $\langle S, \tau\rangle$ has the Bolzano–Weierstrass property and $A \subset S$ is closed, then $\langle A, \tau \mid A\rangle$ has the Bolzano–Weierstrass property.*

Proof. Left to the reader as an exercise.

EXAMPLE 3-13. Let $\langle I^+, \tau\rangle$ be the space described in Example 3-12 and $f: \langle I^+, \tau\rangle \xrightarrow{\text{onto}} \langle I^+, \text{discrete}\rangle$ be given by $f(2n - 1) = f(2n) = n$, $\forall n \in I^+$. Since $f^{-1}(\{n\}) = \{2n - 1, 2n\} \in \tau$, $\forall n \in I^+$, f is continuous. As pointed out in Example 3-12, $\langle I^+, \tau\rangle$ is second countable, has the Bolzano–Weierstrass property, but is not a T_0-space. The image space $\langle I^+, \text{discrete}\rangle$ is a T_5-space that is Lindelöf but does not have the Bolzano–Weierstrass property.

EXAMPLE 3-14. Let $\langle R, \tau\rangle$ be the space described in Example 3-3. We recall that $\tau = \{G \subset R \mid \text{either } 0 \notin G \text{ or } R - \{1, 2\} \subset G\}$. $\langle R, \tau\rangle$ is compact and $\langle R - \{0\}, \tau \mid R - \{0\}\rangle$ is a subspace that is not Lindelöf, since $\tau \mid R - \{0\}$ is the discrete topology. Also, $R - \{0\}$ has no limit in $R - \{0\}$, since the subspace topology is discrete. Thus $\langle R - \{0\}, \tau \mid R - \{0\}\rangle$ does not have the Bolzano–Weierstrass property. Let $\{x_n\}_{n \in I^+}$ be any sequence in $\langle R, \tau\rangle$. If $x_n = 1$ or 2 for at most finitely many values of n, then $x_n \to 0$. If for infinitely many values of n we have $x_n = 1$ (or $x_n = 2$), then some subsequence of $\{x_n\}_{n \in I^+}$ converges to 1 (respectively to 2). In any event, we see that $\langle R, \tau \rangle$ is sequentially compact.

Now we return to the ordinal space $\langle \mathcal{O}, \mathcal{W} \rangle$ discussed earlier in Example 3-7. We show that $\langle \mathcal{O}, \mathcal{W} \rangle$ is first countable, Hausdorff, and sequentially compact but is not a Lindelöf space.

EXAMPLE 3-15. Recall that $\mathcal{W}$ is the order topology on the set $\mathcal{O}$ of ordinals that precede Ω. Let $x \in \mathcal{O}$ and let y be the least element of $\mathcal{O}$ such that $x < y$. Since $x \in \mathcal{O}$, x has only countably many predecessors, say $\{x_n \mid n \in I^+\}$. The collection $\{\{z \in \mathcal{O} \mid x_n < z < y\} \mid n \in I^+\}$ is a countable local base at x for $\mathcal{W}$. Hence $\langle \mathcal{O}, \mathcal{W} \rangle$ is first countable. To see that $\langle \mathcal{O}, \mathcal{W} \rangle$ is Hausdorff, let $x, y \in \mathcal{O}$ with $x < y$. If $\exists z \in \mathcal{O}$ such that $x < z < y$, let $U_x = \{w \in \mathcal{O} \mid w < z\}$ and $V_y = \{w \in \mathcal{O} \mid z < w\}$. If no such z exists, let $U_x = \{w \in \mathcal{O} \mid w < y\}$ and $V_y = \{w \in \mathcal{O} \mid x < w\}$. Then U_x and V_y are disjoint open sets containing x and y, respectively. To see that $\langle \mathcal{O}, \mathcal{W} \rangle$ is not Lindelöf, let G_x be the countable open set $\{w \in \mathcal{O} \mid w < x\}$, $\forall x \in \mathcal{O}$. If $\mathcal{C} = \{G_x \mid x \in \mathcal{O}\}$, then $\mathcal{C}$ is an open cover for $\langle \mathcal{O}, \mathcal{W} \rangle$. If $\mathcal{U} \subset \mathcal{C}$ is countable, then $\bigcup\{G_x \mid G_x \in \mathcal{U}\}$ is countable and cannot contain $\mathcal{O}$, which is uncountable. Since $\langle \mathcal{O}, \mathcal{W} \rangle$ is a first countable T_1-space, we need only show that $\langle \mathcal{O}, \mathcal{W} \rangle$ has the Bolzano–Weierstrass property in order to prove that it is sequentially compact. Let A be an infinite subset of $\mathcal{O}$ and D be a denumerable subset of A. By Theorem 0-3, sup $D \in \mathcal{O}$. Denote by b the least element of $\mathcal{O}$ having infinitely many predecessors in D. If $b \notin D'$, then $\exists c, d \in \mathcal{O}$ such that $c < b < d$ and $\{x \in \mathcal{O} \mid c < x < d\} \cap (D - \{b\}) = \varnothing$. This implies that c has infinitely many predecessors in D, which contradicts the definition of b. Thus $b \in D' \subset A'$ and $\langle \mathcal{O}, \mathcal{W} \rangle$ has the Bolzano–Weierstrass property.

EXERCISES

3-25. Prove Theorem 3-23.

3-26. Prove Theorem 3-24.

3-27. Let $E^1 = \langle R, \xi \rangle$ and $A \subset R$. Show that $\langle A, \xi \mid A \rangle$ has the Bolzano–Weierstrass property iff A is closed and bounded. (This is the classical Bolzano–Weierstrass theorem of real analysis.)

3-28. If $\langle S, \tau_1 \rangle$ and $\langle T, \tau_2 \rangle$ both have the Bolzano–Weierstrass property, does $\langle S \times T, \tau_1 \times \tau_2 \rangle$ have the Bolzano–Weierstrass property? Why or why not?

★3-6 Developable Spaces

At this point in our consideration of the "covering properties," we digress slightly to consider a very important class of topological spaces known as "developable spaces" whose topological structure is obtained from nested sequences of open covers satisfying certain requirements. The metric spaces are a special subclass of the developable spaces, which are a special subclass of the semimetric spaces. The class of developable spaces has been thoroughly studied by R. L. Moore and his students. Indeed, a regular developable space is commonly referred to in the literature as a "Moore space." R. H. Bing has shown that every collectionwise normal Moore space is metrizable. A famous conjecture in topology states that every normal Moore space is collectionwise normal. In Gödel's constructible universe, this conjecture is consistent with set theory. Moreover, if we assume Martin's axiom and deny the Continuum Hypothesis, then there exist normal Moore spaces that are not collectionwise normal.

DEFINITION 3-13. Let $\langle S, \tau \rangle$ be a topological space and $\mathscr{D} = \{\mathscr{G}_n \mid n \in I^+\}$ be a nested sequence of open covers for $\langle S, \tau \rangle$; i.e., $\mathscr{G}_n$ covers $\langle S, \tau \rangle$ and $\mathscr{G}_{n+1}$ is a refinement of $\mathscr{G}_n$, $\forall n \in I^+$. Then $\mathscr{D}$ is a *development* of $\langle S, \tau \rangle$ iff $p \in U \in \tau$ implies that $\exists N \in I^+$ such that $\mathscr{G}_N^*(p) = \bigcup\{G_N \in \mathscr{G}_N \mid p \in G_N\} \subset U$. The set $\mathscr{G}_N^*(p)$ is called the *star of the cover* $\mathscr{G}_N$ *with respect to* p.

DEFINITION 3-14. $\langle S, \tau \rangle$ is a *developable space* iff there exists a development for $\langle S, \tau \rangle$. $\langle S, \tau \rangle$ is a *Moore space* iff $\langle S, \tau \rangle$ is a regular developable T_1-space.

THEOREM 3-25. *Every metric space* $\langle M, d \rangle$ *is a Moore space.*

Proof. $\langle M, d \rangle$ is a T_5-space by Theorem 2-10. Let $\mathscr{G}_n = \{S_d(x; 1/n) \mid x \in M\}$, $\forall n \in I^+$. Then $\mathscr{D} = \{\mathscr{G}_n \mid n \in I^+\}$ is a nested sequence of open covers of $\langle M, d \rangle$. If $x \in U \in \tau_d$, then $\exists N \in I^+$ such that $x \in S_d(x; 1/N) \subset U$. Thus, if $x \in S_d(y; 1/2N)$ and $z \in S_d(y; 1/2N)$, then $d(x, z) \leq d(x, y) + d(y, z) < 1/2N + 1/2N = 1/N$. Thus $S_d(y; 1/2N) \subset S_d(x; 1/N) \subset U$ if $x \in S_d(y; 1/2N) \in \mathscr{G}_{2N}$; i.e., $\mathscr{G}_{2N}^*(x) \subset S_d(x; 1/N) \subset U$. This shows that $\mathscr{D}$ is a development for $\langle M, d \rangle$.

We describe now an example of a Moore space that is not a metric space, since it is not normal—the space of Example 2-4.

EXAMPLE 3-16. $S = \{\langle x, y \rangle \in R \times R \mid y \geq 0\}$. If $\langle p, q \rangle \in S$ and $q > 0$, let $N_\varepsilon(p, q) = \{\langle x, y \rangle \in S \mid (x - p)^2 + (y - q)^2 < \varepsilon^2\}$; for each $\langle p, 0 \rangle \in S$, let $N_\varepsilon(p, 0) = \{\langle x, y\rangle \in S \mid (x - p)^2 + (y - \varepsilon)^2 < \varepsilon^2\} \cup \{\langle p, 0\rangle\}$. Then $\mathscr{B} = \{N_\varepsilon(p, q) \mid \langle p, q\rangle \in S, \varepsilon > 0\}$ is a base for a topology τ on S. As seen in Chapter 2, $\langle S, \tau\rangle$ is a T_3-space but not a T_4-space. Let $\mathscr{G}_n = \{N_{1/n}(p, q) \mid \langle p, q\rangle \in S\}$, $\forall n \in I^+$. Then $\mathscr{D} = \{\mathscr{G}_n \mid n \in I^+\}$ is a nested sequence of open covers of $\langle S, \tau\rangle$. The proof that $\mathscr{D}$ is actually a development for $\langle S, \tau\rangle$ proceeds in much the same way as the proof of Theorem 3-25. Thus we leave the details to the reader as an exercise.

As promised in Section 2-4, we shall now outline Bing's proof of the fact that every metric space is collectionwise normal. To do this, we require some additional notions, "strongly screenable" and "perfectly screenable."

DEFINITION 3-15. $\langle S, \tau\rangle$ is *screenable* iff for each open cover $\mathscr{G}$ of $\langle S, \tau\rangle$ $\exists\{\mathscr{G}_n\}_{n\in I^+}$ such that $\mathscr{G}_n$ is a collection of pairwise disjoint open sets $\forall n \in I^+$ and $\bigcup\{\mathscr{G}_n \mid n \in I^+\} = \{G_n \mid G_n \in \mathscr{G}_n, n \in I^+\}$ is a cover for $\langle S, \tau\rangle$ which is a refinement of $\mathscr{G}$. $\langle S, \tau\rangle$ is *strongly screenable* iff $\exists\{\mathscr{G}_n\}_{n\in I^+}$ as above such that $\mathscr{G}_n$ is discrete $\forall n \in I^+$.

DEFINITION 3-16. $\langle S, \tau\rangle$ is *perfectly screenable* iff there exists a sequence $\{\mathscr{G}_n\}_{n\in I^+}$ such that $\mathscr{G}_n$ is a discrete collection of open sets $\forall n \in I^+$ and such that if $p \in U \in \tau$ $\exists n(p, U) \in I^+$ and $G_n \in \mathscr{G}_{n(p,U)}$ with $p \in G_n \subset U$.

EXAMPLE 3-17. Let $E^1 = \langle R, \xi\rangle$ and $\mathscr{G}$ be an open cover of E^1. Then $\mathscr{G}$ has a countable refinement $\mathscr{U} = \{B_n \mid n \in I^+\}$ where the B_n's are open intervals centered at rational points with rational radii chosen such that $\forall n \in I^+$ $B_n \subset$ some $G \in \mathscr{G}$. Let $\mathscr{G}_n = \{B_n\}$, $\forall n \in I^+$. Then $\{\mathscr{G}_n\}_{n\in I^+}$ has the properties outlined in Definition 3-15, and thus E^1 is strongly screenable. To see that E^1 is also perfectly screenable, let $\mathscr{C}_m = \{(r - 1/m, r + 1/m) \mid r$ rational$\}$, $\forall m \in I^+$. Since E^1 is strongly screenable, for each cover $\mathscr{C}_m$ $\exists\{\mathscr{G}_n^m \mid n \in I^+\}$ such that $\mathscr{G}_n^m$ is a discrete collection of pairwise disjoint open sets $\forall n \in I^+$ and $\bigcup\{\mathscr{G}_n^m \mid n \in I^+\}$ is a cover for E^1 that is a refinement of $\mathscr{C}_m$. Then $\{\mathscr{G}_n^m \mid n \in I^+, m \in I^+\}$ is a countable set of discrete collections of open sets. We may relabel these $\mathscr{G}_1, \mathscr{G}_2, \ldots, \mathscr{G}_k, \ldots$. The reader can easily verify that $\{\mathscr{G}_k\}_{k\in I^+}$ satisfies the requirement of Definition 3-16.

LEMMA (A. H. Stone). *Every metric space $\langle M, d\rangle$ has the property that $\forall \varepsilon > 0$ there exists a sequence $\{\mathscr{F}_{\varepsilon,n}\}_{n\in I^+}$ of discrete collections $\mathscr{F}_{\varepsilon,n}$ of closed subsets of M each of diameter less than ε and $\bigcup\{\mathscr{F}_{\varepsilon,n} \mid n \in I^+\}$ covers M.*

Proof. See A. H. Stone, "Paracompactness and Product Spaces," *Bulletin of the American Mathematical Society* 54(1948), 977–82.

THEOREM 3-26. *Every metric space $\langle M, d\rangle$ is regular and perfectly screenable.*

Proof. By our lemma $\forall \varepsilon > 0\ \exists\{\mathscr{F}_{\varepsilon,n}\}_{n \in I^+}$, where $\mathscr{F}_{\varepsilon,n}$ is a discrete collection of closed subsets of M each of diameter less than ε. For each $F \in \mathscr{F}_{\varepsilon,n}$, let G_F be an open set containing F such that $\delta(G_F) < \varepsilon$ and each point of G_F is more than twice as close to F as to any other set in $\mathscr{F}_{\varepsilon,n}$. Denote by $\mathscr{G}_{\varepsilon,n}$ the collection $\{G_F \mid F \in \mathscr{F}_{\varepsilon,n}\}$, $\forall n \in I^+\ \forall \varepsilon > 0$. The collection $\{\mathscr{G}_{1/m,n} \mid m \in I^+, n \in I^+\}$ is a countable family of discrete collections of open sets. If we relabel them $\mathscr{G}_1, \mathscr{G}_2, \ldots, \mathscr{G}_k, \ldots$, then $\{\mathscr{G}_k\}_{k \in I^+}$ satisfies the requirements of Definition 3-16. Hence $\langle M, d\rangle$ is perfectly screenable.

THEOREM 3-27. *Every perfectly screenable space is strongly screenable.*

Proof. Let $\langle S, \tau\rangle$ be perfectly screenable and $\mathscr{C} = \{U_\alpha \mid \alpha \in \Lambda\}$ any open cover of $\langle S, \tau\rangle$. Then there is a sequence $\{\mathscr{G}_n\}_{n \in I^+}$, where $\mathscr{G}_n$ is a discrete collection of open sets for each $n \in I^+$ and such that if $p \in U \in \tau$, $\exists n(p, U) \in I^+$ and $G_n \in \mathscr{G}_{n(p,U)}$ with $p \in G_n \subset U$. In particular, if $p \in U_\alpha \in \mathscr{C}$, then $\exists n(p, U_\alpha) \in I^+$ and $G_n \in \mathscr{G}_{n(p,U_\alpha)}$ with $p \in G_n \subset U_\alpha$. Let $\mathscr{G}_n' = \{G_n \in \mathscr{G}_{n(p,U_\alpha)} \mid p \in G_n \subset U_\alpha$ whenever $p \in U_\alpha \in \mathscr{C}, \alpha \in \Lambda\}$, $\forall n \in I^+$. It is easily seen that $\mathscr{G}_n' \subset \mathscr{G}_n$ and thus is discrete $\forall n \in I^+$. Also, $\{\mathscr{G}_n'\}_{n \in I^+}$ satisfies the requirements of Definition 3-15. Thus $\langle S, \tau\rangle$ is strongly screenable.

COROLLARY. *Every metric space is regular and strongly screenable.*

THEOREM 3-28. *If $\langle S, \tau\rangle$ is regular and strongly screenable, then $\langle S, \tau\rangle$ is collectionwise normal.*

Proof. Let $\{F_\alpha \mid \alpha \in \Lambda\}$ be any discrete collection of closed subsets of S. Let $\mathscr{C}$ be any open cover of $\langle S, \tau\rangle$ such that the closure of no element of $\mathscr{C}$ intersects two elements of $\{F_\alpha \mid \alpha \in \Lambda\}$. Since $\langle S, \tau\rangle$ is strongly screenable, there is a sequence $\{\mathscr{G}_n\}_{n \in I^+}$ such that $\mathscr{G}_n$ is a discrete collection of open sets and $\bigcup\{\mathscr{G}_n \mid n \in I^+\}$ is a cover for $\langle S, \tau\rangle$ and a refinement of $\mathscr{C}$. Let $U_{n,\beta} = \bigcup\{G_n \in \mathscr{G}_n \mid G_n \cap F_\beta \neq \varnothing\}$ and $V_{n,\beta} = \bigcup\{G_n \in \mathscr{G}_n \mid G_n \cap F_\alpha \neq \varnothing$ for some $\alpha \in \Lambda - \{\beta\}\}$, $\forall \beta \in \Lambda\ \forall n \in I^+$. If $W_\beta = U_{1,\beta} \cup (U_{2,\beta} - \overline{V}_{1,\beta}) \cup \cdots \cup (U_{n,\beta} - \bigcup\{\overline{V}_{i,\beta} \mid i = 1, \ldots, n-1\}) \cup \cdots$, $\forall \beta \in \Lambda$, then $\{W_\beta \mid \beta \in \Lambda\}$ is a collection of pairwise disjoint open sets such that $F_\beta \subset W_\beta$, $\forall \beta \in \Lambda$. Hence $\langle S, \tau\rangle$ is collectionwise normal.

COROLLARY. *Every metric space is collectionwise normal.*

EXERCISES

3-29. Verify that the collection $\mathscr{D}$ described in Example 3-16 is actually a development for the space $\langle S, \tau\rangle$.

3-30. Show that if $\langle S, \tau \rangle$ is developable, then $\langle S, \tau \rangle$ is first countable.

3-31. Show that the property of being developable is hereditary.

3-32. If $\langle S, \tau_1 \rangle$ are $\langle T, \tau_2 \rangle$ are developable, prove that $\langle S \times T, \tau_1 \times \tau_2 \rangle$ is developable. Is the property of being developable countably productive? Why or why not?

3-33. Show that a developable space that has the Lindelöf property is second countable.

3-34. Show that if $\langle S, \tau \rangle$ is compact, then $\langle S, \tau \rangle$ is second countable iff it is developable.

3-35. Prove that a developable space which is both separable and screenable is second countable.

3-36. If a developable space is strongly screenable, is it second countable? Prove or disprove.

★3-7 Paracompactness

We discuss now the property of "paracompactness," which is the last covering property we investigate. A. H. Stone has shown that every metric space is paracompact. Indeed, every regular Lindelöf space is paracompact. E. A. Michael has demonstrated that paracompactness is equivalent to "full normality" for the class of regular spaces. Since R. H. Bing showed that full normality implies collectionwise normality, we are able to conclude that every paracompact regular space is collectionwise normal. In particular, every paracompact Moore space is metrizable. Bing's metrization theorem shows that collectionwise normality implies (hence is equivalent to) paracompactness in any Moore space. Dieudonné proved that a paracompact Hausdorff space is a T_4-space.

DEFINITION 3-17. Let $\langle S, \tau \rangle$ be a topological space and $\mathscr{F} \subset 2^S$. $\mathscr{F}$ is *locally finite* iff $\forall x \in S\ \exists G \in \tau$ such that $x \in G$ and G intersects only finitely many elements of $\mathscr{F}$. $\mathscr{F}$ is *σ-locally finite* iff $\mathscr{F}$ is the union of countably many locally finite families.

DEFINITION 3-18. $\langle S, \tau \rangle$ is *paracompact* iff every open cover for $\langle S, \tau \rangle$ has an open, locally finite refinement.

EXAMPLE 3-18. $\langle R, \text{discrete}\rangle$ is a paracompact Hausdorff space which is neither countably compact nor Lindelöf. $\langle R, \mathscr{L}\rangle$ is also paracompact. For let $\mathscr{C}$ be any open cover for $\langle R, \mathscr{L}\rangle$. For each $r \in R$, $\exists b_r \in R$ such that $r < b_r$ and $[r, b_r)$ is contained in exactly one element of $\mathscr{C}$. Thus the collection $\mathscr{G}$ of these basic open sets $[r, b_r)$ is a cover for $\langle R, \mathscr{L}\rangle$ and a refinement of $\mathscr{C}$. Since $\langle R, \mathscr{L}\rangle$ is a Lindelöf space, $\mathscr{G}$ must contain a countable subcover $\mathscr{G}' = \{[r_n, b_{r_n}) \mid n \in I^+\}$ for $\langle R, \mathscr{L}\rangle$. Let $U_1 = [r_1, b_{r_1})$ and $U_n = [r_n, b_{r_n}) - \bigcup\{[r_i, b_{r_i}] \mid i = 1, \ldots, n-1\}$ for $n > 1$. Then $\mathscr{U} = \{U_n \mid n \in I^+\}$ is an open, locally finite refinement of $\mathscr{C}$. However, $\langle R \times R, \mathscr{L} \times \mathscr{L}\rangle$ is *not* paracompact, since a paracompact Hausdorff space is normal by the theorem of Dieudonné, and we showed earlier that $\langle R \times R, \mathscr{L} \times \mathscr{L}\rangle$ is not normal but is a Hausdorff space. Thus paracompactness is not finitely productive.

Since every finite subcover is a locally finite refinement, it is clear that compactness implies paracompactness. Thus $\langle \mathcal{O}^*, \mathscr{W}^*\rangle$ is paracompact, but the subspace $\langle \mathcal{O}, \mathscr{W}\rangle$ is not paracompact. However, we do have the following analogue of Theorem 3-1, whose proof we leave to the reader as an exercise.

THEOREM 3-29. *If $\langle S, \tau\rangle$ is paracompact and $A \subset S$ is closed, then $\langle A, \tau \mid A\rangle$ is paracompact.*

The following definition of "fully normal" is due to Tukey. Results of E. A. Michael and A. H. Stone imply that fully normal and paracompactness are equivalent. Since the proof is rather lengthy, it will not be given here.

DEFINITION 3-19. Let $\langle S, \tau\rangle$ be a topological space and $\mathscr{C} \subset 2^S$. Then $\mathscr{U} \subset 2^S$ is a *star refinement* of $\mathscr{C}$ iff $\{\mathscr{U}^*(p) \mid p \in S\}$ is a refinement of $\mathscr{C}$, where $\mathscr{U}^*(p) = \bigcup\{U \in \mathscr{U} \mid p \in U\}$ and is called the star of $\mathscr{U}$ with respect to p, $\forall p \in S$. $\langle S, \tau\rangle$ is *fully normal* iff every open cover for $\langle S, \tau\rangle$ has an open star refinement.

We next establish that a regular Lindelöf space is paracompact (Morita's theorem) and that a paracompact Hausdorff space is normal (Dieudonné).

THEOREM 3-30 (Morita). *If $\langle S, \tau\rangle$ is a regular Lindelöf space, then $\langle S, \tau\rangle$ is paracompact.*

Proof. Let $\mathscr{C} = \{G_\alpha \mid a \in \Lambda\}$ be any open cover for $\langle S, \tau\rangle$. By regularity, $\forall x \in S\ \exists \alpha \in \Lambda$ and $U_x, V_x \in \tau$ such that $x \in U_x \subset \overline{U}_x \subset V_x \subset \overline{V}_x \subset G_\alpha$. Thus $\{U_x \mid x \in S\}$ is an open cover for $\langle S, \tau\rangle$ that is a refinement of $\mathscr{C}$. Since $\langle S, \tau\rangle$ is Lindelöf, there is a countable subcover $\{U_{x_n} \mid n \in I^+\}$ for $\langle S, \tau\rangle$.

Let $W_1 = V_{x_1}$ and $W_n = V_{x_n} - \bigcup\{\bar{U}_{x_i} \mid i = 1, \ldots, n-1\}$ $\forall n > 1$. Suppose that $x \in S - \bigcup\{W_n \mid n \in I^+\}$. Let n be the first positive integer for which $x \in V_{x_n}$. Then $x \notin \bigcup\{V_{x_i} \mid i = 1, \ldots, n-1\}$, which implies that $x \notin \bigcup\{\bar{U}_{x_i} \mid i = 1, \ldots, n-1\}$, and hence $x \in W_n$. Contradiction. Thus $\{W_n \mid n \in I^+\}$ is a cover for $\langle S, \tau\rangle$ and is an open refinement of $\mathscr{C}$. If $x \in S$, then $\exists k \in I^+$ such that $x \in U_{x_k}$ and $U_{x_k} \cap W_n = \varnothing$ if $n > k$. Hence $\{W_n \mid n \in I^+\}$ is also locally finite.

THEOREM 3-31 (Dieudonné). *Every paracompact Hausdorff space is normal.*

Proof. Let $\langle S, \tau\rangle$ be any paracompact Hausdorff space, and let $F_1, F_2 \subset S$ be closed and disjoint. Then $\langle F_1, \tau \mid F_1\rangle$ and $\langle F_2, \tau \mid F_2\rangle$ are paracompact by Theorem 3-29. Let $p \in F_1$ be arbitrary but fixed. Since $\langle S, \tau\rangle$ is Hausdorff, for every $q \in F_2$ $\exists U_q, V_q \in \tau$ such that $p \in U_q$, $q \in V_q$, and $U_q \cap V_q = \varnothing$. Then $\mathscr{C} = \{V_q \mid q \in F_2\} \cup \{S - F_2\}$ is an open cover for $\langle S, \tau\rangle$. Since $\langle S, \tau\rangle$ is paracompact, $\mathscr{C}$ has an open, locally finite refinement $\mathscr{C}'$. Let $V_p = \bigcup\{V' \in \mathscr{C}' \mid V' \cap F_2 \neq \varnothing\}$, and U'_p be an open set containing p which intersects only finitely many members of $\mathscr{C}'$, say $V'_1, \ldots, V'_n$. For $i = 1, \ldots, n$, let $V_{q_i} \in \mathscr{C}$ such that $V'_i \subset V_{q_i}$, and let $U_p = U'_p \cap U_{q_1} \cap \cdots \cap U_{q_n}$. Then $p \in U_p$, $F_2 \subset V_p$, and $U_p \cap V_p = \varnothing$. This shows that $\langle S, \tau\rangle$ is regular. Now let $\mathscr{G}$ be the open cover $\{U_p \mid p \in F_1\} \cup \{S - F_1\}$. Then $\mathscr{G}$ has an open, locally finite refinement $\mathscr{G}'$. Let $U = \bigcup\{U' \in \mathscr{G}' \mid U' \cap F_1 \neq \varnothing\}$. Then $F_1 \subset U \in \tau$. For each $q \in F_2$ $\exists W'_q \in \tau$ such that W'_q intersects only finitely many members of $\mathscr{G}'$, say $U'_1, \ldots, U'_m$. For $i = 1, \ldots, m$, let $U_{p_i} \in \mathscr{G}$ such that $U'_i \subset U_{p_i}$, and let $W_q = W'_q \cap V_{p_1} \cap \cdots \cap V_{p_m}$. Let $V = \bigcup\{W_q \mid q \in F_2\}$. Then $V \in \tau$ and $U \cap V = \varnothing$, which establishes the normality of $\langle S, \tau\rangle$.

Finally, we establish Stone's theorem, which states that every metric space is paracompact. The proof is that of Smirnov as given by Pervin in his book *Foundations of Topology*, pp. 162–63.

THEOREM 3-32 (A. H. Stone). *Every metric space $\langle M, d\rangle$ is paracompact.*

Proof. Let $\mathscr{U} = \{U_\alpha \mid \alpha \in \Lambda\}$ be any open cover of $\langle M, d\rangle$ and $<$ a well-ordering for Λ. Let $A \subset M$ and $S_n(A) = \{x \in M \mid d(x, A) < 2^{-n}\}$, $\forall n \in I^+$. Then $A \subset S_n(A) \in \tau_d$ and $S_n(\{x\}) = S_d(x; 2^{-n})$, $\forall n \in I^+$. Let $C_n(A) = \{x \in M \mid S_n(\{x\}) \subset A\}$. Then $C_n(A) = M - S_n(M - A) \subset A$ and is closed. By transfinite induction, we define for each $n \in I^+$ a collection $\{E^n_\alpha \mid \alpha \in \Lambda\}$ of closed subsets of M by letting $E^n_\alpha = C_n(U_\alpha - \bigcup\{E^n_\beta \mid \beta < \alpha\})$, $\forall \alpha \in \Lambda$.

(1) $\{E_\alpha^n \mid \alpha \in \Lambda, n \in I^+\}$ covers M. Let $x \in M$ and λ be the least element of Λ such that $x \in U_\lambda$. Then $\exists n \in I^+$ such that $x \in S_n(\{x\}) = S_d(x; 2^{-n}) \subset U_\lambda$. If $x \notin E_\lambda^n$, then $S_n(\{x\}) \not\subset U_\lambda - \bigcup\{E_\alpha^n \mid \alpha < \lambda\}$. Thus $S_n(\{x\}) \cap (\bigcup\{E_\alpha^n \mid \alpha < \lambda\}) \neq \varnothing$, which implies that $S_n(\{x\}) \cap E_\gamma^n \neq \varnothing$ for some $\gamma < \lambda$. Hence $x \in S_n(E_\gamma^n) = S_n[C_n(U_\gamma - \bigcup\{E_\alpha^n \mid \alpha < \gamma\})] \subset U_\gamma - \bigcup\{E_\alpha^n \mid \alpha < \gamma\} \subset U_\gamma$. Contradiction. Thus $x \in E_\lambda^n$ and $\{E_\alpha^n \mid \alpha \in \Lambda, n \in I^+\}$ covers M.

For each $n \in I^+$ and $\alpha \in \Lambda$, let $F_\alpha^n = \overline{S_{n+3}(E_\alpha^n)}$ and $G_\alpha^n = S_{n+2}(E_\alpha^n)$. Then $F_\alpha^n \subset G_\alpha^n$. If $\alpha < \beta$, then $S_n(E_\beta^n) = S_n[C_n(U_\beta - \bigcup\{E_\alpha^n \mid \alpha < \beta\})] \subset U_\beta - \bigcup\{E_\alpha^n \mid \alpha < \beta\} \subset M - E_\alpha^n$. Hence $S_n(E_\beta^n) \cap E_\alpha^n = \varnothing$, which implies that $d(E_\alpha^n, E_\beta^n) \geq 2^{-n}$ if $\alpha \neq \beta$. This implies that $d(F_\alpha^n, F_\beta^n) \geq 2^{-n} - (2^{-(n+3)} + 2^{-(n+3)}) = 2^{-n} - 2^{-(n+2)} > 2^{-(n+1)}$. Thus $\{F_\alpha^n \mid \alpha \in \Lambda\}$ is discrete and $F^n = \bigcup\{F_\alpha^n \mid \alpha \in \Lambda\}$ is closed, $\forall n \in I^+$. For each $n \in I^+$ and $\alpha \in \Lambda$, let $V_\alpha^n = G_\alpha^n - \bigcup\{F^i \mid i < n\}$. Then V_α^n is open $\forall n \in I^+$, $\forall \alpha \in \Lambda$.

(2) $\{V_\alpha^n \mid \alpha \in \Lambda, n \in I^+\}$ covers M and is a refinement of $\mathscr{U}$. Since $\{E_\alpha^n \mid \alpha \in \Lambda, n \in I^+\}$ covers M, so does $\{F_\alpha^n \mid \alpha \in \Lambda, n \in I^+\}$. If $x \in M$, there exists a least $k \in I^+$ such that $x \in F_\lambda^k$ for some $\lambda \in \Lambda$. Thus $x \in F_\lambda^k - \bigcup\{\bigcup\{F_\alpha^i \mid \alpha \in \Lambda\} \mid i < k\} = F_\lambda^k - \bigcup\{F^i \mid i < k\} \subset G_\lambda^k - \bigcup\{F^i \mid i < k\} = V_\lambda^k$. Since $V_\beta^n \subset G_\beta^n = S_{n+2}(E_\beta^n) \subset S_n(E_\beta^n) = S_n[C_n(U_\beta - \bigcup\{E_\alpha^n \mid \alpha < \beta\})] \subset U_\beta - \bigcup\{E_\alpha^n \mid \alpha < \beta\} \subset U_\beta$, $\{V_\alpha^n \mid \alpha \in \Lambda, n \in I^+\}$ is a refinement of $\mathscr{U}$.

(3) $\{V_\alpha^n \mid \alpha \in \Lambda, n \in I^+\}$ is locally finite. If $x \in M$, then $x \in E_\lambda^k$ for some $\lambda \in \Lambda$ and $k \in I^+$. Then $S_{k+3}(\{x\}) \subset S_{k+3}(E_\lambda^k) \subset \overline{S_{k+3}(E_\lambda^k)} = F_\lambda^k \subset F^k$, which implies that $S_{k+3}(\{x\}) \cap V_\alpha^n = \varnothing$, $\forall n > k$, $\forall \alpha \in \Lambda$. If $n \leq k$, then $d(F_\alpha^n, F_\gamma^n) \geq 2^{-(k+1)}$, which implies that $d(E_\alpha^n, E_\gamma^n) > 2^{-(k+1)}$. However, $\sup\{d(y, z) \mid y, z \in S_{k+3}(\{x\})\} \leq 2^{-(k+2)} < 2^{-(k+1)}$, which implies that $S_{k+3}(\{x\})$ is an open set containing x which intersects at most one member of $\{E_\alpha^n \mid \alpha \in \Lambda\}$ for each $n \leq k$. Hence $S_{k+3}(\{x\})$ intersects at most k members of $\{E_\alpha^n \mid \alpha \in \Lambda, n \in I^+\}$ and thus at most k members of $\{V_\alpha^n \mid \alpha \in \Lambda, n \in I^+\}$.

EXERCISES

3-37. Prove Theorem 3-29.

3-38. $\langle S, \tau\rangle$ is *countably paracompact* iff every countable open cover for $\langle S, \tau\rangle$ has an open, locally finite refinement. Show that any closed subspace of a countably paracompact space is countably paracompact.

3-39. Show that if every open subspace of a paracompact space is paracompact, then the space is hereditarily paracompact.

3-40. Prove that if $\langle S, \tau_1\rangle$ is paracompact and $\langle T, \tau_2\rangle$ is compact, then $\langle S \times T, \tau_1 \times \tau_2\rangle$ is paracompact.

3-41. Let $\langle S, \tau \rangle$ be a topological space and $\mathscr{F} \subset 2^S$ be locally finite (discrete). Show that $\overline{\mathscr{F}} = \{\bar{F} \mid F \in \mathscr{F}\}$ is also locally finite (respectively, discrete), and that $\overline{\bigcup\{F \mid F \in \mathscr{F}\}} = \bigcup\{\bar{F} \mid F \in \mathscr{F}\}$.

3-42. $\langle S, \tau \rangle$ is *σ-compact* iff $\langle S, \tau \rangle$ is the union of countably many compact subspaces. Prove that every σ-compact space has the Lindelöf property.

3-43. Show that every F_σ-subspace of a paracompact space is paracompact.

★3-8 Compactification

We conclude this chapter with a brief discussion of some ways in which we can topologically embed a noncompact space in a compact space under certain circumstances. We consider here only the two best-known compactifications, the one-point compactification of Alexandroff and the Stone–Čech compactification. The first method applies to the class of locally compact spaces, and the second method applies to the class of completely regular spaces.

DEFINITION 3-20. A *compactification* of a space $\langle S, \tau \rangle$ is a pair $\langle \langle \hat{S}, \hat{\tau} \rangle, h \rangle$, where $\langle \hat{S}, \hat{\tau} \rangle$ is a compact Hausdorff space and h is a homeomorphism of $\langle S, \tau \rangle$ onto a dense subspace of $\langle \hat{S}, \hat{\tau} \rangle$.

DEFINITION 3-21. Let $\langle S, \tau \rangle$ be a locally compact, noncompact Hausdorff space and $p \notin S$. Let $\hat{S} = S \cup \{p\}$ and $\hat{\tau} = \tau \cup \{G \subset \hat{S} \mid \hat{S} - G$ is a compact subset of $S\}$. Then $\langle \hat{S}, \hat{\tau} \rangle$ is compact Hausdorff. If h is the identity homeomorphism of $\langle S, \tau \rangle$ onto itself, then $\langle \langle \hat{S}, \hat{\tau} \rangle, h \rangle$ is the *one-point* (*Alexandroff*) *compactification* of $\langle S, \tau \rangle$.

EXAMPLE 3-19. The one-point compactification of $E^1 = \langle R, \xi \rangle$ (which is locally compact) is $\langle \langle R \cup \{\infty\}, \hat{\xi} \rangle$, identity mapping$\rangle$, where $\hat{\xi} = \xi \cup \{G \subset R \cup \{\infty\} \mid R \cup \{\infty\} - G \subset R$ and compact$\}$. Thus $G = \{x \in R \cup \{\infty\} \mid x < a$ or $x > b\}$ is an open set containing ∞. The one-point compactification of $\langle (0, 1], \xi \mid (0, 1] \rangle$ is $\langle \langle [0, 1], \xi \mid [0, 1] \rangle$, identity mapping$\rangle$.

THEOREM 3-33 (Alexandroff). *Any locally compact, noncompact Hausdorff space $\langle S, \tau \rangle$ can be uniquely embedded in a compact Hausdorff space $\langle \hat{S}, \hat{\tau} \rangle$ such that $\hat{S} - S$ is a singleton.*

Proof. Let $p \notin S$ and $\hat{S} = S \cup \{p\}$. Let $\hat{\tau} = \tau \cup \{G \subset \hat{S} \mid \hat{S} - G$ is a compact subset of $S\}$. The reader can easily check that $\hat{\tau}$ is a topology on $\hat{S}$, since it satisfies axioms (O1)–(O3). Clearly, the identity homeomorphism establishes $\langle S, \tau \rangle$ as a dense subspace of $\langle \hat{S}, \hat{\tau} \rangle$. By construction $\hat{S} - S$ is a singleton. Thus it remains to show that $\langle \hat{S}, \hat{\tau} \rangle$ is compact Hausdorff and unique. Let $x \in S$. Since $\langle S, \tau \rangle$ is locally compact Hausdorff $\exists G \in \tau$ and compact $K \subset S$ such that $x \in G \subset \bar{G} \subset K$. Thus $\bar{G}$ is compact, which implies that $p \in S - \bar{G} \in \hat{\tau}$. Clearly, $G \cap (S - \bar{G}) = \varnothing$ and $\langle \hat{S}, \hat{\tau} \rangle$ is Hausdorff. To see that $\langle \hat{S}, \hat{\tau} \rangle$ is compact, let $\mathscr{U} = \{G_\alpha \mid \alpha \in \Lambda\}$ be any open cover for $\langle \hat{S}, \hat{\tau} \rangle$. Then $p \in G_\beta$ for some $\beta \in \Lambda$. Hence $\hat{S} - G_\beta \subset S$ and compact. Thus $\{G_\alpha \mid \alpha \in \Lambda - \{\beta\}\}$ is an open cover for $\hat{S} - G_\beta$ and some finite subcollection $\{G_{\alpha_i} \mid i = 1, \ldots, n\}$ also covers $\hat{S} - G_\beta$. Then $\{G_{\alpha_i} \mid i = 1, \ldots, n\} \cup \{G_\beta\}$ is a finite subcover of $\mathscr{U}$ for $\langle \hat{S}, \hat{\tau} \rangle$. Suppose that $\langle \hat{T}, \hat{\sigma} \rangle$ is also a one-point compactification of $\langle S, \tau \rangle$; i.e., $\hat{T} - S = \{q\}$, where $q \notin S$ and $\hat{\sigma} \mid S = \tau$. Let $f(x) = x$, $\forall x \in S$, and $f(p) = q$. Then $f: \langle \hat{S}, \hat{\tau} \rangle \to \langle \hat{T}, \hat{\sigma} \rangle$ is bijective. If $G \in \tau$, then $f(G) = G \in \hat{\sigma}$. Also, if $p \in \hat{S} - C \in \hat{\tau}$, then $C = f(C) \subset S$ is compact and $q = f(p) \in f(\hat{S} - C) = \hat{T} - C \in \hat{\sigma}$. Thus f is open. A similar argument shows that f^{-1} is open, and hence f is a homeomorphism.

DEFINITION 3-22. Let $\langle S, \tau \rangle$ be a completely regular Hausdorff space. If $I^S = \{f: S \to I^1 \mid f$ is bounded and continuous$\}$ and $I_f = I^1$, $\forall f \in I^S$, then $\prod\{I_f \mid f \in I^S\}$ is a compact Tychonoff space and the evaluation map $e: S \to \prod\{I_f \mid f \in I^S\}$ given by $[e(x)]_f = f(x)$ is an embedding. The *Stone–Čech compactification* of $\langle S, \tau \rangle$ is $\langle \beta(S), e \rangle$, where $\beta(S) = \overline{e(S)}$.

EXAMPLE 3-20. $[0, 1]$ is not the Stone–Čech compactification of $(0, 1]$ since the continuous function $f(x) = \sin(1/x)$ on $(0, 1]$ does *not* have a continuous extension over $[0, 1]$. Similarly, $[-1, 1]$ is not the Stone–Čech compactification of $(-1, 1)$. Moreover, the unit circle S^1 is not the Stone–Čech compactification of E^1, since the continuous function $f(x) = \arctan x$ on E^1 cannot be extended continuously to S^1. However, if P denotes the Tychonoff plank (Example 2-6), then P is the Stone–Čech compactification of $P - \{\langle \Omega, \omega \rangle\}$, since each continuous real-valued function on $P - \{\langle \Omega, \omega \rangle\}$ can be extended to P. For more details the reader is referred to Dugundji's *Topology* and Willard's *General Topology*.

THEOREM 3-34 (Stone–Čech)

(1) *If $\langle T, \tau_2 \rangle$ is compact and if $f: \langle S, \tau_1 \rangle \to \langle T, \tau_2 \rangle$ is continuous, then there is a unique continuous $F: \beta(S) \to T$ such that $f = F \circ e$.*

(2) *Any compactification $\langle\langle \hat{S}, \hat{\tau}_1 \rangle, h \rangle$ of $\langle S, \tau_1 \rangle$ having property* (1) *is homeomorphic to $\beta(S)$.*

(3) *$\beta(S)$ is the "largest" compactification of $\langle S, \tau_1\rangle$; i.e., if $\langle \hat{S}, \hat{\tau}_1\rangle$ is any compactification of $\langle S, \tau_1\rangle$, then $\langle \hat{S}, \hat{\tau}_1\rangle$ is a quotient space of $\beta(S)$.*

Proof

(1) The following diagram commutes:

$$\begin{array}{ccc} S & \xrightarrow{f} & T \\ {\scriptstyle e}\downarrow & & \downarrow{\scriptstyle e_0} \\ \beta(S) & \xrightarrow[\varphi]{} & \beta(T) \end{array}$$

Since $\langle T, \tau_2\rangle$ is compact, $\beta(T)$ is homeomorphic to T. Let $F = e_0^{-1} \circ \varphi$. Then $F: \beta(S) \to T$ is continuous and $f = e_0^{-1} \circ \varphi \circ e = F \circ e$. Moreover, F is unique, since $\langle S, \tau\rangle$ is densely embedded in $\beta(S)$ by e.

(2) S may be regarded as a subset of both $\hat{S}$ and $\beta(S)$. Let $i: S \to S$ be the identity mapping. Then (1) implies that there is a unique $F: \beta(S) \to \hat{S}$ such that $F \mid S = i$ and there is a unique $G: \hat{S} \to \beta(S)$ such that $G \mid S = i^{-1}$. Thus $F \circ G \mid S = i$ and $G \circ F \mid S = i$, and S is a dense subset of both $\hat{S}$ and $\beta(S)$. Hence $F \circ G$ is the identity mapping on $\hat{S}$ and $G \circ F$ is the identity mapping on $\beta(S)$. Thus F and G are homeomorphisms.

(3) The identity mapping $i: S \to S$ has a continuous extension $F: \beta(S) \to \hat{S}$ by (1). Since $\beta(S)$ is compact, $F[\beta(S)]$ is compact and thus is a closed set containing the dense subset S of $\hat{S}$. This implies that F is onto. Since F is also closed, $\hat{S}$ is homeomorphic to $\beta(S)/K(F)$, where $K(F)$ is the equivalence relation on $\beta(S)$ given by $\langle x, y\rangle \in K(F)$ iff $F(x) = F(y)$.

EXERCISES

3-44. Describe the one-point compactification of $E^2 = E^1 \times E^1$.

3-45. Prove that every locally compact Hausdorff space is completely regular.

3-46. Let $\langle S, \tau\rangle$ be a locally compact Hausdorff space, and let $C \subset S$ be compact. If $C \subset G \in \tau$, show that there exists a continuous function $f: S \to [0, 1]$ such that $f(C) = \{1\}$ and $f(S - G) = \{0\}$.

CHAPTER 4
Connectivity Properties

4-1 Connectivity

In contrast to the separation axioms, we now investigate "connectivity properties." Intuitively, a space is connected iff it consists of a single piece ("component") relative to the topology. Also, a space S is pathwise (arcwise) connected iff each two points of the space can be joined by a path (arc) in S. The connectivity of E^1 assures its completeness.

DEFINITION 4-1. $\langle S, \tau\rangle$ is *connected* iff $S \neq G_1 \cup G_2$, where $G_1, G_2 \in \tau - \{\varnothing\}$ and $G_1 \cap G_2 = \varnothing$. $\langle S, \tau\rangle$ is *disconnected* iff $\langle S, \tau\rangle$ is not connected.

The above definition implies that $\langle S, \tau\rangle$ is connected iff no proper subset of S is both open and closed. Moreover, if $\langle C, \tau \mid C\rangle$ is a subspace of $\langle S, \tau\rangle$ that is connected and C is contained in the union of two disjoint open sets, then it is contained in one of them.

EXAMPLE 4-1. Every subspace of $\langle S, \text{indiscrete}\rangle$ is connected.

EXAMPLE 4-2. $\langle S, \text{discrete}\rangle$ is totally disconnected; i.e., the singletons are the largest connected subspaces.

EXAMPLE 4-3. $\langle R, \xi\rangle$ and $\langle R, \text{cofinite}\rangle$ are connected. However, $\langle R, \mathscr{L}\rangle$ is totally disconnected, since

$$[a, b) = \left[a, \frac{a+b}{2}\right) \cup \left[\frac{a+b}{2}, b\right) \qquad \forall a, b \in R.$$

It is easily seen that connectivity is not hereditary, since the removal of the midpoint of the (connected) interval (a, b) in E^1 separates it into two intervals,

$$\left(a, \frac{a+b}{2}\right) \quad \text{and} \quad \left(\frac{a+b}{2}, b\right).$$

However, connectivity does have an important "interpolation" property described in the following theorem. In particular, the closure of a connected subspace is connected.

THEOREM 4-1. *Let $\langle S, \tau\rangle$ be a topological space and $\langle C, \tau \mid C$ be a$\rangle$ connected subspace. If $C \subset D \subset \bar{C}$, then $\langle D, \tau \mid D\rangle$ is connected.*

Proof. Suppose that $\langle D, \tau \mid D\rangle$ were not connected. Then $\exists D \cap G_1, D \cap G_2 \in \tau \mid D - \{\varnothing\}$ such that $D = (D \cap G_1) \cup (D \cap G_2) = D \cap (G_1 \cup G_2)$ and $D \cap G_1 \cap G_2 = \varnothing$. Since $C \subset D$ and $\langle C, \tau \mid C\rangle$ is connected, either $C \subset D \cap G_1$ or $C \subset D \cap G_2$. If $C \subset D \cap G_1$, then $\varnothing \neq D \cap G_2 \subset C'$. This implies that $(D \cap G_2) \cap C \neq \varnothing$ and $D \cap G_1 \cap G_2 \neq \varnothing$. Contradiction. The case $C \subset D \cap G_2$ leads to the same contradiction. Thus $\langle D, \tau \mid D\rangle$ is connected.

THEOREM 4-2. *Let $\langle C_\alpha, \tau_\alpha\rangle$ be a connected subspace of $\langle S, \tau\rangle$, $\forall \alpha \in \Lambda$, and $C_\beta \cap C_\gamma \neq \varnothing$, $\forall \beta, \gamma \in \Lambda$. If $C = \bigcup\{C_\alpha \mid \alpha \in \Lambda\}$, then $\langle C, \tau \mid C\rangle$ is connected.*

Proof. Suppose that $\langle C, \tau \mid C\rangle$ were not connected. Then $\exists C \cap G_1, C \cap G_2 \in \tau \mid C - \{\varnothing\}$ such that $C = (C \cap G_1) \cup (C \cap G_2)$ and $C \cap G_1 \cap G_2 = \varnothing$. Since $\langle C_\alpha, \tau_\alpha\rangle$ is connected, $\forall \alpha \in \Lambda$ and $C_\beta \cap C_\gamma \neq \varnothing$, $\forall_\beta, \gamma \in \Lambda$, we must have either $C_\alpha \subset C \cap G_1$, $\forall \alpha \in \Lambda$, or $C_\alpha \subset C \cap G_2$, $\forall \alpha \in \Lambda$. This implies that either $C \subset C \cap G_1$ (hence $C \cap G_2 = \varnothing$) or $C \subset C \cap G_2$ (hence $C \cap G_1 = \varnothing$). Contradiction in either case. Thus $\langle C, \tau \mid C\rangle$ is connected.

Next, we show that connectivity is a continuous invariant and that the graph of a continuous function with a connected domain is connected.

THEOREM 4-3. *If $f: \langle S, \tau_1\rangle \xrightarrow{\text{onto}} \langle T, \tau_2\rangle$ is continuous and $\langle S, \tau_1\rangle$ is connected, then $\langle T, \tau_2\rangle$ is connected.*

Proof. Suppose that $\langle T, \tau_2\rangle$ is not connected. Then $\exists G_1, G_2 \in \tau_2 - \{\varnothing\}$ such that $G_1 \cap G_2 = \varnothing$ and $T = G_1 \cup G_2$. Since f is continuous, $f^{-1}(G_1)$, $f^{-1}(G_2) \in \tau_1 - \{\varnothing\}$. Also, $S = f^{-1}G_1) \cup f^{-1}(G_2)$ and $f^{-1}(G_1) \cap f^{-1}(G_2) = \varnothing$. Hence $\langle S, \tau_1\rangle$ is not connected. Contradiction.

THEOREM 4-4. *If $\langle S, \tau_1\rangle$ is connected and $f: \langle S, \tau_1\rangle \to \langle T, \tau_2\rangle$ is continuous, then the graph of f is connected with respect to the product topology.*

Proof. Let G denote the graph of f and suppose that $\langle G, \tau_1 \times \tau_2 \mid G\rangle$ is not connected. Then $\exists \varnothing \neq A \subset G$ such that $A, G - A \in \tau_1 \times \tau_2 \mid G$. Since the projection $\pi_1 : S \times T \xrightarrow{\text{onto}} S$ is continuous and open, $\pi_1 \mid G : G \xrightarrow{\text{onto}} S$ is open. Thus $[\pi_1 \mid G](A) \in \tau_1 - \{\varnothing\}$ and $[\pi_1 \mid G](G - A) \in \tau_1 - \{\varnothing\}$. Also $S = [\pi_1 \mid G](A) \cup [\pi_1 \mid G](G - A)$ and $[\pi_1 \mid G](A) \cap [\pi_1 \mid G](G - A) = \varnothing$. Thus $\langle S, \tau_1\rangle$ is not connected. Contradiction.

In Chapter 3, we proved the Heine–Borel–Lebesgue theorem, which characterizes the compact subspaces of E^1. We now obtain a characterization of the connected subspaces of E^1 using the following lemma.

LEMMA. *A subspace $\langle A, \xi \mid A \rangle$ of E^1 is connected iff whenever $a, b, c \in R$ with $a, b \in A$ and $a < c < b$, then $c \in A$.*

Proof. Let $\langle A, \xi \mid A \rangle$ be connected and $a, b, c \in R$ with $a, b \in A$ and $a < c < b$. Suppose that $c \notin A$. Then $G_1 = \{x \in A \mid x < c\}$ and $G_2 = \{x \in A \mid c < x\}$ are nonempty, disjoint, (relatively) open sets whose union is A. This implies that $\langle A, \xi \mid A \rangle$ is not connected. Contradiction. *Conversely,* suppose that $\langle A, \xi \mid A \rangle$ is a subspace of E^1 for which the condition of the theorem holds and yet $\langle A, \xi \mid A \rangle$ is not connected. Then $\exists G_1, G_2 \in \xi \mid A - \{\varnothing\}$ such that $G_1 \cap G_2 = \varnothing$ and $A = G_1 \cup G_2$. Without loss of generality, we may suppose the labeling is such that $\exists a \in G_1, b \in G_2$, with $a < b$. Let $G_1^* = \{x \in R \mid [a, x] \cap G_2 = \varnothing\}$. By the completeness property of E^1, G_1^* has a least upper bound c, since it has an upper bound b. Since $a \leq c \leq b$, either $c \in G_1$ or $c \in G_2$. If $c \in G_1$, there is an open interval containing c that has an empty intersection with G_2. Contradicts the statement that c is an upper bound for G_1^*. Thus $c \in G_2$ and some open interval about c has an empty intersection with G_1. This implies that $\exists d \in R$ such that $a < d < c \leq b$, $d \notin G_1$, and $d \in A = G_1 \cup G_2$. Also, $d \notin G_2$, since $\exists e \in G_1^*$ such that $d < e < c$ by definition of c as the least upper bound of G_1^* and thus $[a, d] \cap G_2 = \varnothing$, since $[a, e] \cap G_2 = \varnothing$. Contradiction.

THEOREM 4-5. *A subspace of $E^1 = \langle R, \xi \rangle$ is connected iff it is one of the following types:*

(1) *$\varnothing$ or R.*

(2) *$\{x\}$, where $x \in R$.*

(3) *(a, b), $[a, b)$, $(a, b]$, or $[a, b]$, where $a, b \in R$ with $a < b$.*

(4) *$\{x \mid x < a\}$, $\{x \mid x \leq a\}$, $\{x \mid x > a\}$, or $\{x \mid x \geq a\}$, where $a \in R$.*

Proof. Follows from the lemma above. Details left to the reader as an exercise.

An important consequence of Theorem 4-5 is the famous "intermediate-value theorem," which follows. The proof is left to the reader as an exercise.

THEOREM 4-6. *If $f: \langle [a, b], \xi \mid [a, b] \rangle \to E^1$ is continuous and $f(a) < \gamma < f(b)$, then there is c such that $a < c < b$ and $f(c) = \gamma$.*

We conclude this section with a proof of the fact that connectivity is a productive property. As a consequence, we have that E^n and I^n are connected.

LEMMA. *Let $x^0 \in \prod_\Lambda S_\alpha$, where $\prod_\Lambda S_\alpha$ has the Tychonoff product topology $\prod_\Lambda \tau_\alpha$. If $D = \{x \in \prod_\Lambda S_\alpha \mid x(\alpha) \neq x^0(\alpha)$ for only finitely many $\alpha \in \Lambda\}$, then D is dense in $\prod_\Lambda S_\alpha$.*

Proof. Let $B = \bigcap\{\pi_{\alpha_i}^{-1}(G_{\alpha_i}) \mid i = 1, \ldots, n\}$ be any basic open set in $\prod_\Lambda S_\alpha$. Then $\exists y \in B$ such that $y(\alpha) = x^0(\alpha)$, $\forall \alpha \in \Lambda - \{\alpha_1, \ldots, \alpha_n\}$. Thus $y \in D$ and D is dense in $\prod_\Lambda S_\alpha$.

THEOREM 4-7. *If $\langle S_\alpha, \tau_\alpha\rangle$ is connected $\forall \alpha \in \Lambda$, then $\langle \prod_\Lambda S_\alpha, \prod_\Lambda \tau_\alpha\rangle$ is connected.*

Proof. We show by induction that if $x^0_{(n)}(\alpha) \neq x^0(\alpha)$ for at most n values of $\alpha \in \Lambda$, then $x^0_{(n)}$ and x^0 lie in a connected subset of $\prod_\Lambda S_\alpha$. This is true for $n = 1$, since if $x^0_{(1)}(\alpha) \neq x^0(\alpha)$, the slice of the product space through x^0 parallel to $\langle S_\alpha, \tau_\alpha\rangle$ is homeomorphic to $\langle S_\alpha, \tau_\alpha\rangle$, and thus is a connected subspace containing $x^0_{(1)}$ and x^0. Assume that the statement is true for all $x^0_{(n-1)} \in \prod_\Lambda S_\alpha$. For an arbitrary $x^0_{(n)} \in \prod_\Lambda S_\alpha$, let $x^0_{(n-1)}$ be a point that differs from $x^0_{(n)}$ in only one coordinate. There is a connected subset C_1 containing $x^0_{(n)}$ and $x^0_{(n-1)}$. By our induction hypothesis, there is a connected subset C_2 containing $x^0_{(n-1)}$ and x^0. Then $C_1 \cup C_2$ is a connected subset containing $x^0_{(n)}$ and x^0. Let C be the union of all connected subsets of $\prod_\Lambda S_\alpha$ containing x^0. C is connected by Theorem 4-2 and contains D, which is dense in $\prod_\Lambda S_\alpha$ by our lemma. Hence $D \subset \bar{D} = \prod_\Lambda S_\alpha \subset C \subset \prod_\Lambda S_\alpha$, which implies that $\prod_\Lambda S_\alpha = C$.

EXERCISES

4-1. Show that $\langle S, \tau\rangle$ is connected iff no proper subset of S is both open and closed.

4-2. Let $\langle S, \tau\rangle$ be a topological space and $A, B \in 2^S$. A and B are *separated* in $\langle S, \tau\rangle$ iff $\bar{A} \cap B = \varnothing = A \cap \bar{B}$. Prove that $\langle S, \tau\rangle$ is connected iff $\langle S, \tau\rangle$ is not the union of two nonempty sets that are separated in $\langle S, \tau\rangle$.

4-3. Prove Theorem 4-5.

4-4. Prove Theorem 4-6.

4-5. Show that $E^n - \{p\}$ is connected $\forall p \in E^n$ if $n > 1$, and use this to show that $E^n - F$ is connected for every finite subset F of E^n if $n > 1$.

4-6. Prove that if $f\colon I^1 \to I^1$ is continuous, then $\exists x \in I^1$ such that $f(x) = x$ (such an x is called a *fixed point*).

4-7. Show that if $\langle M, d\rangle$ is a connected metric space containing more than one point, then M is uncountable.

4-8. $\langle S, \tau\rangle$ is *strongly connected* (*Steen and Seebach*) iff there do not exist any nonconstant continuous mappings $f: \langle S, \tau\rangle \to E^1$. Show that if $\langle S, \tau\rangle$ is strongly connected (Steen and Seebach), then $\langle S, \tau\rangle$ is connected. Give an example of a connected space that is *not* strongly connected (Steen and Seebach).

4-9. $\langle S, \tau\rangle$ is *strongly connected* (*Levine*) iff whenever $S \subset G_1 \cup G_2$ with $G_1, G_2 \in \tau$, then either $S \subset G_1$ or $S \subset G_2$. Show that if $\langle S, \tau\rangle$ is strongly connected (Levine), then $\langle S, \tau\rangle$ is strongly connected (Steen and Seebach). Give an example of a space that is strongly connected (Steen and Seebach) but is *not* strongly connected (Levine).

4-2 Components and Continua

In this section we investigate two types of connected subspaces of a topological space. The first type is called a "component" and is a maximal connected subspace. The second type is a subspace that is both compact and connected, called a "continuum."

DEFINITION 4-2. $\langle C, \tau \mid C\rangle$ is a *component* of $\langle S, \tau\rangle$ iff $\langle C, \tau \mid C\rangle$ is connected and is not a proper subspace of any other connected subspace of $\langle S, \tau\rangle$.

DEFINITION 4-3. Let $[x] = \{y \in S \mid S$ is not the union of two disjoint open sets, one containing x and the other $y\}$. Then $\{[x] \mid x \in S$ is the collection of *quasicomponents* of $\langle S, \tau\rangle$.

EXAMPLE 4-4. The components of $\langle R, \text{discrete}\rangle$ and $\langle R, \mathscr{L}\rangle$ are the singletons. Thus both spaces are totally disconnected. $\langle R, \xi\rangle$, $\langle R, \text{cofinite}\rangle$, and $\langle R, \text{indiscrete}\rangle$ are connected and thus each has a single component (itself).

EXAMPLE 4-5. Let $S = I^+$ and τ have as a base the collection of sets $B_n = \{2n - 1, 2n\}$, where $n \in I^+$. Then the components of $\langle S, \tau\rangle$ are the doubleton sets B_n, where $n \in I^+$.

EXAMPLE 4-6. Let $\langle S, \tau\rangle$ be the subspace of E^2 shown in Figure 4-1. The components of $\langle S, \tau\rangle$ are the individual line segments $y = 1/n$ ($|x| \leq 1$),

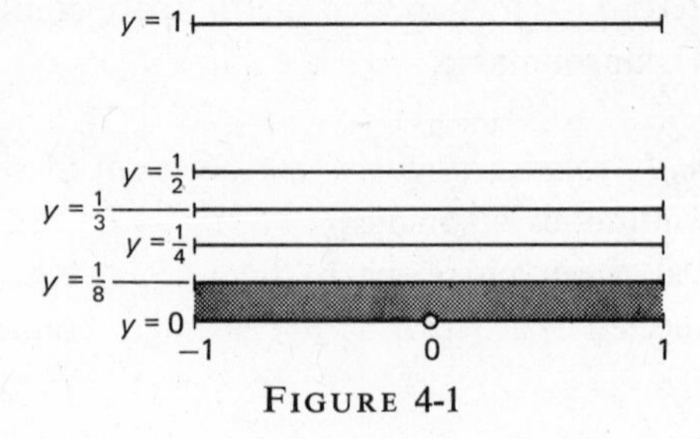

FIGURE 4-1

$\forall n \in I^+$, and the intervals $[-1, 0)$ and $(0, 1]$ of the set $Q = \{\langle x, 0\rangle \mid 0 < |x| \leq 1\}$. The set Q is the limit set in $\langle S, \tau\rangle$ of the sequence of components $y = 1/n$ $(|x| \leq 1)$. Q is a quasicomponent but not a component, since Q is not connected.

This example shows that a quasicomponent may not be a component. However, we show that each component of a space lies in a quasicomponent of the space, that the components and quasicomponents are closed, and that the components of a space partition the space.

THEOREM 4-8. *If $\langle S, \tau\rangle$ is a topological space and $x \in S$, there is a component C and a quasicomponent Q of S such that $x \in C \subset Q$.*

Proof. Let $C = \{y \in S \mid x$ and y lie in some connected subspace of $\langle S, \tau\rangle\}$. C is connected by Theorem 4-2 and is maximal by construction. Thus C is a component containing x. Let $Q = \{y \in S \mid S \neq G_1 \cup G_2$ with $G_1, G_2 \in \tau - \{\varnothing\}, G_1 \cap G_2 = \varnothing$, and $x \in G_1, y \in G_2\}$. $C \subset Q$ by virtue of Definition 4-1, and Q is a quasicomponent by Definition 4-3.

THEOREM 4-9. *Each component and quasicomponent of $\langle S, \tau\rangle$ is closed.*

Proof. Let $\langle C, \tau \mid C\rangle$ be any component of $\langle S, \tau\rangle$. Then $\langle \bar{C}, \tau \mid \bar{C}\rangle$ is connected, since $\langle C, \tau \mid C\rangle$ is connected. By the maximality of $\langle C, \tau \mid C\rangle$ we must have $\bar{C} = C$. Hence C is closed. Now let $\langle Q, \tau \mid Q\rangle$ be any quasicomponent of $\langle S, \tau\rangle$, and let $p \in Q'$. If $S = G_1 \cup G_2$, where $G_1, G_2 \in \tau$ and $G_1 \cap G_2 = \varnothing$, then $p \in G_1$ or $p \in G_2$ (not both). If $p \in G_1$, then $\exists x \in Q \cap G_1$ such that $x \neq p$. This implies that $Q \subset G_1$, and thus $p \in Q$. The argument is entirely similar in the case $p \in G_2$. Hence, $Q' \subset Q$ and Q is closed.

THEOREM 4-10. *The components of $\langle S, \tau\rangle$ form a partition of $\langle S, \tau\rangle$.*

Proof. By Theorem 4-8 each point x of S lies in that component which is the union of all connected subsets of S containing x. Suppose now that

$\langle C_1, \tau \mid C_1\rangle$ and $\langle C_2, \tau \mid C_2\rangle$ are components of $\langle S, \tau\rangle$ and that $C_1 \cap C_2 \neq \varnothing$. Then $\exists p \in C_1 \cap C_2$. Since C_2 is a connected set containing $p \in C_1$, we have $C_2 \subset C_1$ by the maximality of C_1. Also, C_1 is a connected set containing $p \in C_2$, which implies that $C_1 \subset C_2$ by the maximality of C_2. Hence $C_1 = C_2$ and the components form a partition of $\langle S, \tau\rangle$.

We now introduce the concept of a "continuum," which is a compact connected space. $\langle R, \text{cofinite}\rangle$ and $\langle R, \text{indiscrete}\rangle$ are continua. However, $\langle R, \xi\rangle$, $\langle R, \mathscr{L}\rangle$, and $\langle R, \text{discrete}\rangle$ are not continua, because they are not compact, and $\langle R, \mathscr{L}\rangle$ and $\langle R, \text{discrete}\rangle$ are not even connected. Also, the spaces described in Examples 4-5 and 4-6 are not continua, since they are neither compact nor connected. The subspace I^1 of E^1 is a continuum, as is any homeomorphic image of I^1 (called an "arc").

DEFINITION 4-4. $\langle S, \tau\rangle$ is a *continuum* iff $\langle S, \tau\rangle$ is compact and connected.

DEFINITION 4-5. If $\langle S, \tau\rangle$ is connected and $p \in S$, then p is a *cut point* of S iff $\langle S - \{p\}, \tau \mid (S - \{p\})\rangle$ is disconnected. Otherwise, p is a *noncut point* of S.

In terms of the above definitions we have the following characterizations of the "arc" and the "simple closed curve," which we merely state without proofs. A metric continuum $\langle M, d\rangle$ is an arc iff M has exactly two noncut points. A metric continuum $\langle M, d\rangle$ is a simple closed (Jordan) curve iff for each two points $x, y \in M$, the subspace $\langle M - \{x, y\}, d\rangle$ is disconnected.

EXERCISES

4-10. Let $\langle x, y\rangle \in R$ iff x and y lie in a connected subspace of $\langle S, \tau\rangle$. Show that R is an equivalence relation on S and describe the equivalence classes.

4-11. Let $\langle C, \tau \mid C\rangle$ be a connected subspace of $\langle S, \tau\rangle$ and suppose that C is both open and closed in S. Show that $\langle C, \tau \mid C\rangle$ is a component of $\langle S, \tau\rangle$.

4-12. Is the property of being a component a continuous invariant? Why or why not?

4-13. If $\langle C_1, \tau_1 \mid C_1\rangle$ is a component of $\langle S, \tau_1\rangle$ and $\langle C_2, \tau_2 \mid C_2\rangle$ is a component of $\langle T, \tau_2\rangle$, is $\langle C_1 \times C_2, \tau_1 \times \tau_2 \mid C_1 \times C_2\rangle$ necessarily a component of $\langle S \times T, \tau_1 \times \tau_2\rangle$? Support your answer.

4-14. Prove that the quasicomponent of $\langle S, \tau \rangle$ containing $x \in S$ is the intersection of all subspaces of S that contain x and are both open and closed.

4-15. Is the property of being a continuum a continuous invariant? Why or why not?

4-16. Is the property of being a continuum productive? Support your answer.

4-17. Is the property of being a cut point (noncut point) a continuous invariant? Support your answer in each case.

4-18. Let $f: \langle S, \tau_1 \rangle \xrightarrow{\text{onto}} \langle T, \tau_2 \rangle$ be continuous and $\langle T, \tau_2 \rangle$ connected. If p is a cut point of $\langle T, \tau_2 \rangle$, show that $S - \{f^{-1}(p)\}$ is disconnected.

4-3 Local Connectivity

In much the same way as we "localized" compactness in Exercise 3-7, we can "localize" connectivity. We investigate the property of a space $\langle S, \tau \rangle$ being "locally connected at $p \in S$" and the slightly weaker property of $\langle S, \tau \rangle$ being "connected *im kleinen* at $p \in S$."

DEFINITION 4-6. $\langle S, \tau \rangle$ is *locally connected at* $p \in S$ iff whenever $p \in U \in \tau$, there exists a connected $V \in \tau$ such that $p \in V \subset U$. $\langle S, \tau \rangle$ is *locally connected* iff $\langle S, \tau \rangle$ is locally connected at p for each $p \in S$.

As seen earlier, every compact space is also locally compact. In contrast, a connected space need not be locally connected. Also, a space may be locally connected at all of its points except one and still be connected, as the following example shows.

EXAMPLE 4-7. Let G denote the graph of $y = \sin(1/x)$, where $0 < x \leq 1$ with the subspace Euclidean topology $\xi \times \xi \mid G$ (see Figure 4-2). $\langle G, \xi \times \xi \mid G \rangle$ is connected by Theorem 4-4 and is also easily seen to be locally connected. Let $S = G \cup \{\langle 0, 0 \rangle\}$. Then $\langle S, \xi \times \xi \mid S \rangle$ is connected by Theorem 4-1 but fails to be locally connected at $\langle 0, 0 \rangle$. Let $T = G \cup \{\langle 0, y \rangle \mid -1 \leq y \leq 1\}$. Then $\langle T, \xi \times \xi \mid S \rangle$ is a continuum (called the topologist's sine curve), which fails to be locally connected at $\langle 0, y \rangle$ $\forall y \in [-1, 1]$.

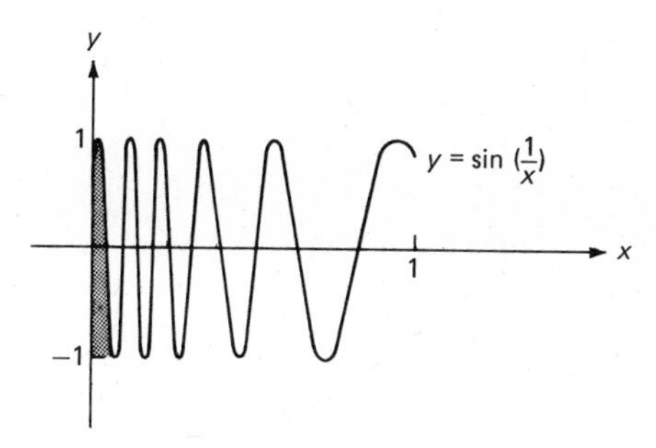

FIGURE 4-2

EXAMPLE 4-8. E^1 is locally connected, as are $\langle R$, discrete$\rangle$ and $\langle R$, cofinite$\rangle$. However, $\langle R, \mathscr{L}\rangle$ is not locally connected at any point, since no basic set $[a, b)$ is connected. Also, if A denotes the set of rational numbers, then the subspace $\langle A, \xi \mid A\rangle$ of E is not locally connected since no member of $\xi \mid A - \{\varnothing\}$ is connected. This shows that the property of being locally connected is not hereditary.

Open subspaces of a locally connected space are locally connected, and local connectivity is invariant under continuous open mappings, as the reader may easily demonstrate. Next, we give an example to show that local connectivity is not a continuous invariant and obtain a characterization of a locally connected space in terms of components.

THEOREM 4-11. *If $\langle S, \tau\rangle$ is locally connected and $A \in \tau$, then $\langle A, \tau \mid A\rangle$ is locally connected.*

Proof. Left to the reader as an exercise.

THEOREM 4-12. *If $f: \langle S, \tau_1\rangle \xrightarrow{onto} \langle T, \tau_2\rangle$ is continuous and open, and if $\langle S, \tau_1\rangle$ is locally connected, then $\langle T, \tau_2\rangle$ is locally connected.*

Proof. Left to the reader as an exercise.

EXAMPLE 4-9. Let $S = I^+ \cup \{0\}$ have the discrete topology. Let $T = \{1/n \mid n \in I^+\} \cup \{0\}$ have the Euclidean subspace topology $\xi \mid T$. Then $\langle S$, discrete$\rangle$ is locally connected and $\langle T, \xi \mid T\rangle$ is not locally connected. However, the function $f: S \to T$ given by $f(0) = 0$ and $f(n) = 1/n$ is a continuous bijection.

THEOREM 4-13. *$\langle S, \tau\rangle$ is locally connected iff each component of an open subset of S is open.*

Proof. Let $\langle S, \tau \rangle$ be locally connected and $A \in \tau - \{\varnothing\}$. If $\langle C, \tau \mid C \rangle$ is any component of $\langle A, \tau \mid A \rangle$, then $\forall x \in C$ there is a connected $U_x \in \tau$ such that $x \in U_x \subset A$, since $\langle A, \tau \mid A \rangle$ is locally connected by Theorem 4-11. Hence $C \cup U_x$ is a connected subset of A containing x so that $U_x \subset C$ by the maximality of C. This implies that $C = \bigcup \{U_x \mid x \in C\}$ and thus is open. Conversely, suppose that each component of each open subset of S is open. Let $x \in S$ and $x \in U \in \tau$. Then $\langle U, \tau \mid U \rangle$ is an open subspace of $\langle S, \tau \rangle$. Each point $y \in U$ lies in a component C_y of U, and C_y is open by our hypothesis. In particular, $x \in C_x \subset U$ and thus $\langle S, \tau \rangle$ is locally connected.

COROLLARY. *Any component of any locally connected space is both open and closed.*

We give now an example to show that local connectivity is not a productive property. Then, as an exercise, the reader is asked to show that the product of two (hence finitely many) locally connected spaces is locally connected.

EXAMPLE 4-10. Let $\langle S_n, \tau_n \rangle$ be the space $\langle \{0, 1\}, \text{discrete} \rangle$, $\forall n \in I^+$, and let $\langle S, \tau \rangle = \langle \prod_{I^+} S_n, \prod_{I^+} \tau_n \rangle$. Let $x(n) = 0$, $\forall n \in I^+$. Then $x \in S$, but no element of τ containing x is connected. Hence $\langle S, \tau \rangle$ is not locally connected at x.

Briefly, we investigate the property "connected *im kleinen* at p," which is slightly weaker than "locally connected at p" as we show by example. Also, we prove that a space which is connected *im kleinen* at every point is locally connected.

DEFINITION 4-7. $\langle S, \tau \rangle$ is *connected im kleinen at* $x \in S$ iff whenever $x \in U \in \tau$ $\exists V \in \tau$ such that $x \in V \subset U$ and if $y \in V$, there is a connected subset of U containing $\{x, y\}$. $\langle S, \tau \rangle$ is *connected im kleinen* iff $\langle S, \tau \rangle$ is connected *im kleinen* at x for each $x \in S$.

EXAMPLE 4-11. The space shown in Figure 4-3 is connected *im kleinen* at x but is not locally connected at x.

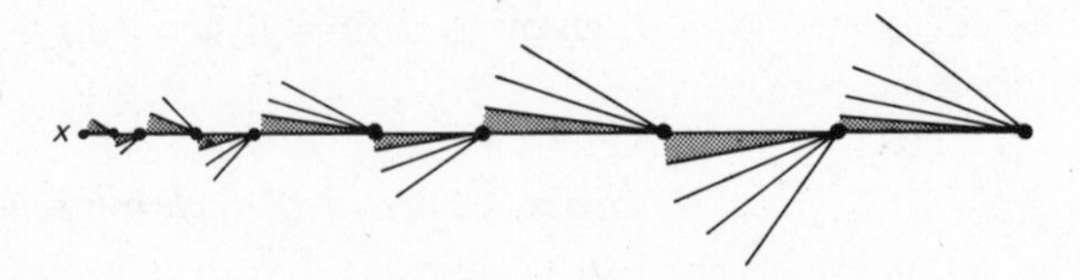

FIGURE 4-3

THEOREM 4-14. *If $\langle S, \tau \rangle$ is connected im kleinen, then it is locally connected.*

Proof. Let $U \in \tau - \{\varnothing\}$ and $\langle C, \tau \mid C \rangle$ be any component of $\langle U, \tau \mid U \rangle$. If $x \in C$, then $\exists V_x \in \tau$ such that $x \in V_x \subset U$, and if $y \in V_x$, there is a connected subset C_y of U such that $\{x, y\} \subset C_y \subset C$. This implies that $V_x \subset C$ so that $C = \bigcup\{V_x \mid x \in C\}$. Hence $C \in \tau$, and $\langle S, \tau \rangle$ is locally connected by Theorem 4-13.

EXERCISES

4-19. Show that $\langle S, \tau \rangle$ is locally connected iff τ has a base consisting of connected open sets.

4-20. Prove Theorem 4-11.

4-21. Prove Theorem 4-12.

4-22. Show that if $\langle S, \tau_1 \rangle$ and $\langle T, \tau_2 \rangle$ are locally connected, then $\langle S \times T, \tau_1 \times \tau_2 \rangle$ is locally connected.

4-23. Show that $\langle \prod_\Lambda S_\alpha, \prod_\Lambda \tau_\alpha \rangle$ is locally connected iff $\langle S_\alpha, \tau_\alpha \rangle$ is locally connected $\forall \alpha \in \Lambda$ and $\langle S_\alpha, \tau_\alpha \rangle$ is connected for all but finitely many $\alpha \in \Lambda$.

4-24. Show that if $f: \langle S, \tau_1 \rangle \xrightarrow{\text{onto}} \langle T, \tau_2 \rangle$ is continuous and open and if $\langle S, \tau_1 \rangle$ is connected *im kleinen*, then $\langle T, \tau_2 \rangle$ is connected *im kleinen*.

4-25. If $\langle S, \tau_1 \rangle$ and $\langle T, \tau_2 \rangle$ are connected *im kleinen* at $p \in S$ and $q \in T$, respectively, is $\langle S \times T, \tau_1 \times \tau_2 \rangle$ connected *im kleinen* at $\langle p, q \rangle$?

4-26. A metric space $\langle M, d \rangle$ has *property S* iff M is the union of a finite number of connected sets of diameter less than ε for each $\varepsilon > 0$. Prove that if $\langle M, d \rangle$ has property S, then $\langle M, d \rangle$ is connected *im kleinen* and thus is locally connected.

4-27. Prove that if a metric space $\langle M, d \rangle$ is compact and locally connected, then $\langle M, d \rangle$ has property S.

★4-4 Pathwise Connectivity

A "curve" in a space $\langle S, \tau \rangle$ is understood to be any subspace of $\langle S, \tau \rangle$ that is the image of I^1 under continuous mapping. The mapping f itself is called a "path" in $\langle S, \tau \rangle$ from $f(0)$ to $f(1)$. $\langle S, \tau \rangle$ is "pathwise connected" iff each two points of S can be joined by a path in $\langle S, \tau \rangle$. We investigate here the property of pathwise connectivity, which we shall use extensively in our study of homotopy theory in Chapter 7.

DEFINITION 4-8. If $\langle S, \tau \rangle$ is a topological space and $p, q \in S$, then a *path in* $\langle S, \tau \rangle$ *from* p *to* q is a continuous function $f: I^1 \to \langle S, \tau \rangle$ such that $f(0) = p$ and $f(1) = q$.

DEFINITION 4-9. $\langle S, \tau \rangle$ is *pathwise connected* iff each two points of S can be joined by a path in $\langle S, \tau \rangle$.

EXAMPLE 4-12. $\langle S$, indiscrete$\rangle$ is always pathwise connected, and $\langle S$, discrete$\rangle$ is never pathwise connected if S contains two or more points.

EXAMPLE 4-13. Let $S = \{0, 1\}$ and $\tau = \{\varnothing, \{0\}, S\}$. Then $\langle S, \tau \rangle$ is called the "Sierpinski space." It is pathwise connected, since the function $f: I^1 \to \langle S, \tau \rangle$ given by

$$f(t) = \begin{cases} 0 & \text{if } t \in [0, 1) \\ 1 & \text{if } t = 1 \end{cases}$$

is continuous and thus a path joining 0 and 1.

EXAMPLE 4-14. $S^n = \{\langle x_1, x_2, \ldots, x_{n+1} \rangle \in E^{n+1} \mid \sum_{k=1}^{n+1} x_k^2 = 1\}$ is called the n-sphere and is the cover of the unit ball $B^{n+1} = \{\langle x_1, x_2, \ldots, x_{n+1} \rangle \in E^{n+1} \mid \sum_{k=1}^{n+1} x_k^2 \leq 1\}$ in E^{n+1}. The reader can easily see that both E^n and $S^n (n \geq 1)$ are pathwise connected, since in either space any two points are the endpoints of an arc (topological copy of I^1) in that space, which provides us with a path that is even a homeomorphism.

We obtain the following alternative characterization of pathwise connectivity, which we shall find useful later in our study of homotopy theory.

THEOREM 4-15. *Let* $x_0 \in S$. *Then* $\langle S, \tau \rangle$ *is pathwise connected iff each* $x \in S$ *can be joined to* x_0 *by a path in* $\langle S, \tau \rangle$. (In this case we call x_0 a "base point" for the collection of paths in $\langle S, \tau \rangle$.)

Proof. The necessity of the condition is clear from Definition 4-9. Conversely, suppose that x_0 is a base point for the collection of paths in $\langle S, \tau \rangle$. If $x_1, x_2 \in S$, then $\exists f_1 : I^1 \to \langle S, \tau \rangle$ and $f_2 : I^1 \to \langle S, \tau \rangle$ such that $f_1(0) = x_1$, $f_1(1) = x_0 = f_2(0)$, and $f_2(1) = x_2$. Then function $g : I^1 \to \langle S, \tau \rangle$ given by

$$g(t) = \begin{cases} f_1(2t) & \text{if } t \in [0, \frac{1}{2}] \\ f_2(2t - 1) & \text{if } t \in [\frac{1}{2}, 1] \end{cases}$$

is continuous (why?), and $g(0) = f_1(0) = x_1$ and $g(1) = f_2(1) = x_2$. Hence g is a path from x_1 to x_2 in $\langle S, \tau \rangle$ and $\langle S, \tau \rangle$ is pathwise connected.

Pathwise connectivity implies connectivity, but a space may be connected without being pathwise connected. This is shown by the topologist's sine curve (Example 4-7), which is connected, not locally connected, and not pathwise connected (why not?).

THEOREM 4-16. *If $\langle S, \tau \rangle$ is pathwise connected, then $\langle S, \tau \rangle$ is connected.*

Proof. Follows from Theorems 4-15 and 4-2. Details are left to the reader as an exercise.

The graph of $y = \sin(1/x)$ $(0 < x \leq 1)$ is clearly pathwise connected. However, as remarked above, its closure, which we call the topologist's sine curve, is not pathwise connected. The reader is invited to show that pathwise connectivity is a continuous invariant and that any family of pathwise-connected spaces with a point in common has a union that is also pathwise connected.

THEOREM 4-17. *If $f: \langle S, \tau_1 \rangle \xrightarrow{onto} \langle T, \tau_2 \rangle$ is continuous and $\langle S, \tau_1 \rangle$ is pathwise connected, then $\langle T, \tau_2 \rangle$ is pathwise connected.*

THEOREM 4-18. *If $\langle A_\alpha, \tau_\alpha \rangle$ is a pathwise-connected subspace of $\langle S, \tau \rangle$, $\forall \alpha \in \Lambda$, and if $\bigcap\{A_\alpha \mid \alpha \in \Lambda\} \neq \varnothing$, then $\langle A, \tau \mid A \rangle$ is pathwise connected where $A = \bigcup\{A_\alpha \mid \alpha \in \Lambda\}$.*

DEFINITION 4-10. The *path components* of $\langle S, \tau \rangle$ are the maximal pathwise-connected subspaces of $\langle S, \tau \rangle$.

Observe that the path components of $\langle S, \tau \rangle$ constitute a partition of $\langle S, \tau \rangle$ but that a path component need not be closed. In particular, the graph of $y = \sin(1/x)$ $(0 < x \leq 1)$ is a path component of the topologist's sine curve but is not closed.

The question arises as to what topological condition must be added to connectivity in order to get pathwise connectivity. This question is answered by the results that conclude this section.

THEOREM 4-19. *Each path component of $\langle S, \tau \rangle$ is open (hence also closed) iff for each $x \in S$ $\exists G_x \in \tau$ such that $x \in G_x$ and $\langle G_x, \tau \mid G_x \rangle$ is pathwise connected.*

Proof. If each path component is open, then simply let G be the path component containing x for each $x \in S$. Conversely, suppose that each $x \in S$ lies in a pathwise-connected open set G_x. Let C be any path component of $\langle S, \tau \rangle$. Since C is a maximal pathwise-connected set containing x for each $x \in C$, we have $x \in G_x \subset C$, which implies that $C = \bigcup\{G_x \mid x \in C\}$ is open. Since $S - C$ is the union of the other path components and thus is open, we must have that C is closed.

THEOREM 4-20. *$\langle S, \tau \rangle$ is pathwise connected iff $\langle S, \tau \rangle$ is connected and for each $x \in S$ $\exists G_x \in \tau$ such that $\langle G_x, \tau \mid G_x \rangle$ is pathwise connected.*

Proof. If $\langle S, \tau \rangle$ is pathwise connected, then $x \in S \in \tau$, $\forall x \in S$, and $\langle S, \tau \rangle$ is connected by Theorem 4-16. Conversely, suppose that $\langle S, \tau \rangle$ is connected and each $x \in S$ lies in a pathwise-connected open subspace of $\langle S, \tau \rangle$. By Theorem 4-19, each path component of $\langle S, \tau \rangle$ is both open and closed in $\langle S, \tau \rangle$. Since $\langle S, \tau \rangle$ is connected, no proper subset of S is both open and closed in S. Hence $\langle S, \tau \rangle$ must have only a single path component (itself) and so is pathwise connected.

COROLLARY. *An open subspace of E^n (or S^n) is connected iff it is pathwise connected.*

EXERCISES

4-28. Show that if $f: I^1 \to \langle S, \tau \rangle$ is a path from x to y in $\langle S, \tau \rangle$, then $g: I^1 \to \langle S, \tau \rangle$ given by $g(t) = f(1 - t)$ is a path from y to x in $\langle S, \tau \rangle$.

4-29. Supply the details in the proof of Theorem 4-16.

4-30. Prove Theorem 4-17.

4-31. Prove Theorem 4-18.

4-32. Prove that the path components of $\langle S, \tau \rangle$ constitute a partition of $\langle S, \tau \rangle$.

4-33. Show that if either condition of Theorem 4-19 holds, then the path components of $\langle S, \tau \rangle$ are identical with the components of $\langle S, \tau \rangle$.

4-34. Show that $\langle \prod_\Lambda S_\alpha, \prod_\Lambda \tau_\alpha \rangle$ is pathwise connected iff $\langle S_\alpha, \tau_\alpha \rangle$ is pathwise connected $\forall \alpha \in \Lambda$.

★4-5 Arcwise Connectivity

An "arc" in $\langle S, \tau \rangle$ is any subspace of $\langle S, \tau \rangle$ that is homeomorphic to I^1. We say that $\langle S, \tau \rangle$ is "arcwise connected" if each two points of S can be joined by an arc in $\langle S, \tau \rangle$. In this section we discuss the property of a space being arcwise connected and the related notion of a "Peano space," which is a locally connected metric continuum.

DEFINITION 4-11. A subspace $\langle A, \tau \mid A \rangle$ of $\langle S, \tau \rangle$ is an *arc* in $\langle S, \tau \rangle$ iff $\langle A, \tau \mid A \rangle$ is homeomorphic to I^1.

Recall the characterization, given in Section 4-2, of the arc. An arc is a metric continuum with exactly two noncut points.

DEFINITION 4-12. $\langle S, \tau \rangle$ is *arcwise connected* iff each two points of S can be joined by an arc in $\langle S, \tau \rangle$.

We show that every arcwise-connected space is connected. Also, we give an example of a connected space that is not arcwise connected.

EXAMPLE 4-15. $\langle R, \text{discrete} \rangle$ and $\langle R, \mathscr{L} \rangle$ cannot be arcwise connected, since they are totally disconnected. $E^1 = \langle R, \xi \rangle$ is arcwise connected, however.

EXAMPLE 4-16. The space $\langle G, \xi \times \xi \mid G \rangle$ described in Example 4-7 is arcwise connected. However, the other connected spaces $\langle S, \xi \times \xi \mid S \rangle$ and $\langle T, \xi \times \xi \mid T \rangle$ described in that same example, and having $\langle G, \xi \times \xi \mid G \rangle$ as a proper subspace, are not arcwise connected.

THEOREM 4-21. *If $\langle S, \tau \rangle$ is arcwise connected, then $\langle S, \tau \rangle$ is pathwise connected.*

Proof. Let $x, y \in S$. There is an arc $\langle A, \tau \mid A \rangle$ joining x and y in $\langle S, \tau \rangle$. Hence there is a homeomorphism $h: I^1 \xrightarrow{\text{onto}} A$ such that $h(0) = x$ and $h(1) = y$. By Definition 4-8, h is a path joining x and y. Thus $\langle S, \tau \rangle$ is pathwise connected.

COROLLARY. *If $\langle S, \tau \rangle$ is arcwise connected, then $\langle S, \tau \rangle$ is connected.*

Proof. Immediate consequence of Theorems 4-21 and 4-16.

Finally, we define the concept of a "Peano space," which is a special type of arcwise-connected metric space.

DEFINITION 4-13. A metric space $\langle M, d \rangle$ is a *Peano space* iff $\langle M, d \rangle$ is a locally connected continuum.

THEOREM 4-22. *Every Peano space* $\langle M, d \rangle$ *is arcwise connected.*

Proof. (Outline only; for details, see Hocking and Young, *Topology*, pp. 116–17.) Let $a, b \in M$ with $a \neq b$. There exists a "nested" sequence $\{K_n\}_{n \in I^+}$ of continua in $\langle M, d \rangle$ such that $\{a, b\} \subset K_n$ and $K_{n+1} \subset K_n$, $\forall n \in I^+$. Let $K = \bigcap\{K_n \mid n \in I^+\}$. One shows that if $x \in K - \{a, b\}$, then x is a cut point of K. Then using our characterization of the arc, we conclude that K is an arc joining a and b, since a and b are the only two noncut points. Thus $\langle M, d \rangle$ is arcwise connected.

EXERCISES

4-35. Is arcwise connectivity a continuous invariant? Why or why not?

4-36. Is the product of two arcwise-connected spaces arcwise connected? Prove or disprove.

4-37. The *arc components* of $\langle S, \tau \rangle$ are the maximal arcwise-connected subspaces of $\langle S, \tau \rangle$. Are the arc components of $\langle S, \tau \rangle$ closed? Do the arc components of $\langle S, \tau \rangle$ constitute a partition of $\langle S, \tau \rangle$? Support your answers. Also, show that the arc component containing x is a subspace of the path component containing x, which is a subspace of the component containing x, $\forall x \in S$.

4-38. Is the property of being a Peano space a continuous invariant? Why or why not?

4-39. Is the product of two Peano spaces a Peano space? Prove or disprove.

4-40. Reconsider Exercises 4-35, 4-36, and 4-37 in the case in which all the spaces discussed are also required to be Hausdorff spaces.

CHAPTER 5

Metrization

5-1 Metrizability

One of the classical problems of topology involves finding conditions on a topological space $\langle S, \tau \rangle$ such that one can define a metric d on $S \times S$ such that the metric topology τ_d induced on S by d is identically τ. The problem was partially solved by Urysohn in 1924 and completely solved independently by Nagata, Smirnov, and Bing in 1950. Their results are presented in this section. Two generalizations of metrizability, semimetrizability and a-metrizability, are considered in this chapter. Also discussed are uniform spaces and topological groups.

DEFINITION 5-1. A topological space $\langle S, \tau \rangle$ is *metrizable* iff there is a metric d on S such that $\tau_d = \tau$.

From Definition 5-1 it follows that $\langle S, \tau \rangle$ is metrizable iff there exists a metric d on S such that $\{S_d(x; r) \mid r > 0\}$ is a local base at x for each $x \in S$. Details are left to the reader as an exercise.

Our first result is Urysohn's metrization theorem, which states that every second countable T_3-space is metrizable. The proof involves topologically embedding such a space in the Hilbert cube I^ω, which we now define.

DEFINITION 5-2. The *Hilbert cube* I^ω is the metric space $\langle S, d \rangle$, where $S = \prod\{[0, 1/n] \mid n \in I^+\}$ and

$$d[(x_1, x_2, \ldots), (y_1, y_2, \ldots)] = \left[\sum_{i=1}^{\infty} (x_i - y_i)^2\right]^{1/2}.$$

THEOREM 5-1 (Urysohn). *If $\langle S, \tau \rangle$ is a second countable T_3-space, then $\langle S, \tau \rangle$ is metrizable.*

Proof. Let $\mathscr{B} = \{B_n \mid n \in I^+\}$ be any countable base for τ. For each $j \in I^+$, $\exists x \in B_j$ and $G \in \tau$ such that $x \in G \subset \bar{G} \subset B_j$, since $\langle S, \tau \rangle$ is regular. Since $\mathscr{B}$ is a base for τ, $\exists i \in I^+$ such that $x \in B_i \subset G \subset \bar{G} \subset B_j$, which implies that $x \in B_i \subset \overline{B_i} \subset B_j$. The collection of pairs $\langle B_i, B_j \rangle$ having the property just described is denumerable and thus can be put in 1-1 correspondence with I^+. Let $\langle B_i, B_j \rangle$ be the nth pair. By Theorem 3-11, $\langle S, \tau \rangle$ is normal (T_4), since it is regular (T_3) and Lindelöf. Since $\overline{B_i}$ and $S - B_j$ are disjoint closed sets, there is a continuous function $f_n : \langle S, \tau \rangle \to I^1$ such that

$f_n(\overline{B_i}) = \{0\}$ and $f_n(S - B_j) = \{1\}$ by Urysohn's lemma (Theorem 2-7). Define a mapping $f: \langle S, \tau \rangle \to I^\omega$ by $f(x) = \langle 2^{-1}f_1(x), 2^{-2}f_2(x), \ldots, 2^{-n}f_n(x), \ldots \rangle$. The rest of the proof involves showing f is a homeomorphism onto $f(S)$.

Let $x, y \in S$ with $x \neq y$. Then $\exists B_j \in \mathscr{B}$ such that $x \in B_j$, $y \in S - B_j$, since $\langle S, \tau \rangle$ is a T_1-space. Then $\exists B_i \in \mathscr{B}$ such $x \in B_i \subset \overline{B_i} \subset B_j$. The nth coordinate of $f(x)$ is thus $2^{-n}f_n(x) = 0$, since $x \in \overline{B_i}$. The nth coordinate of $f(y)$ is $2^{-n}f_n(y) = 2^{-n} \neq 0$, since $y \in S - B_j$. This shows that f is injective, since $f(x) \neq f(y)$.

Let $x_0 \in S$ and $\varepsilon > 0$. Then $\exists N \in I^+$ such that $\sum_{n=N+1}^{\infty} 2^{-2n} < \varepsilon^2/2$. For each n such that $1 \leq n \leq N$, f_n is continuous and thus $\exists B_n \in \mathscr{B}$ such that $x_0 \in B_n$ and $|f_n(x_0) - f_n(x)| < \varepsilon/\sqrt{2N}$ if $x \in B_n$. Then $B = \bigcap\{B_n \mid 1 \leq n \leq N\} \in \tau$ and contains x_0. If $x \in B$, then $x \in B_n$ for each n such that $1 \leq n \leq N$, and thus

$$\begin{aligned}
d(f(x_0), f(x)) &= \left[\sum_{n=1}^{\infty} [2^{-n}f_n(x_0) - 2^{-n}f_n(x)]^2\right]^{1/2} \\
&= \left[\sum_{n=1}^{N} 2^{-2n}|f_n(x_0) - f_n(x)|^2 + \sum_{n=N+1}^{\infty} 2^{-2n}|f_n(x_0) - f_n(x)|^2\right]^{1/2} \\
&\leq \left[\sum_{n=1}^{N} |f_n(x_0) - f_n(x)|^2 + \sum_{n=N+1}^{\infty} 2^{-2n}\right]^{1/2} \\
&< \left[N\left(\frac{\varepsilon}{\sqrt{2N}}\right)^2 + \frac{\varepsilon^2}{2}\right]^{1/2} = \varepsilon.
\end{aligned}$$

This shows that f is continuous at x_0 and hence on S, since x_0 was arbitrary.

It remains to show that f is open. Let $G \in \tau$ and $y = \langle y_1, y_2, \ldots \rangle \in f(G)$. Then $\exists x \in G$ such that $y = f(x)$. Also, for some $n \in I^+$, the nth pair $\langle B_i, B_j \rangle$ has the property that $x \in B_i \subset \overline{B_i} \subset B_j \subset G$. Thus $f_n(x) = 0$ and $f_n(S - G) = \{1\}$. If $p \in S - G$, then $d(f(x), f(p)) \geq 2^{-n}$, which implies that $f(S - G) \cap S_d(f(x); 2^{-n}) = \varnothing$. Hence $y \in S_d(y; 2^{-n}) \cap f(S) \subset f(G)$, which implies that $f(G)$ is open relative to $f(S)$.

Our next results are the metrization theorems of Nagata, Smirnov, and Bing. The theorem of Nagata and Smirnov states that a topological space $\langle S, \tau \rangle$ is metrizable iff it is a T_3-space with a σ-locally finite base. Its proof involves topologically embedding such a space in the generalized Hilbert space H^α of weight α. We define H^α and prove some lemmas before proving the main theorem.

DEFINITION 5-3. Let Λ be an index set of infinite cardinality α. If H^α is the set of all real-valued mappings f defined on Λ such that each mapping is identically zero except on a countable subset of Λ and $\sum_{\lambda \in \Lambda} [f(\lambda)]^2 < \infty$, and d^α is a metric on H^α given by $d^\alpha(f, g) = [\sum_{\lambda \in \Lambda} [f(\lambda) - g(\lambda)]^2]^{1/2}$, then $\langle H^\alpha, d^\alpha \rangle$ is the *generalized Hilbert space of weight* α.

LEMMA 5-1. *If* $\langle S, \tau \rangle$ *is a* T_3*-space with a* σ*-locally finite base, then each member of* τ *is an* F_σ *set.*

Proof. Let $\mathscr{B} = \{B_{n,\lambda} \mid \lambda \in \Lambda_n, n \in I^+\}$ be a σ-locally finite base for τ and $G \in \tau$. By regularity, $\exists B_{n(x),\lambda(x)} \in \mathscr{B}$ such that $x \in B_{n(x),\lambda(x)} \subset \overline{B}_{n(x),\lambda(x)} \subset G$ for each $x \in G$. Let $B_n = \bigcup\{B_{n,\lambda(x)} \mid x \in G\}$. Then $\overline{B_n} = \bigcup\{\overline{B_{n,\lambda(x)}} \mid x \in G\} \subset G$ by local finiteness and Exercise 3-41. This implies that $G = \bigcup\{\overline{B_n} \mid n \in I^+\}$ and thus is an F_σ.

LEMMA 5-2. *If* $\langle S, \tau \rangle$ *is a* T_3*-space with a* σ*-locally finite base, then* $\langle S, \tau \rangle$ *is normal.*

Proof. Let $\mathscr{B} = \{B_{n,\lambda} \mid \lambda \in \Lambda_n, n \in I^+\}$ be a σ-locally finite base for τ, and let F_1 and F_2 be disjoint closed subsets of S. By regularity for each $x \in F_1$, $\exists B_{n(x),\lambda(x)} \in \mathscr{B}$ such that $x \in B_{n(x),\lambda(x)} \subset \overline{B}_{n(x),\lambda(x)} \subset S - F_2$. Similarly, for each $y \in F_2$, $\exists B_{n(y),\lambda(y)} \in \mathscr{B}$ such that $y \in B_{n(y),\lambda(y)} \subset \overline{B}_{n(y),\lambda(y)} \subset S - F_1$. Let $B_{n,F_1} = \bigcup\{B_{n,\lambda(x)} \mid x \in F_1\}$ and $B_{n,F_2} = \bigcup\{B_{n,\lambda(y)} \mid y \in F_2\}$. By local finiteness and Exercise 3-41, we have $\overline{B}_{n,F_1} = \bigcup\{\overline{B}_{n,\lambda(x)} \mid x \in F_1\} \subset S - F_2$ and $\overline{B}_{n,F_2} = \bigcup\{\overline{B}_{n,\lambda(y)} \mid y \in F_2\} \subset S - F_1$. The sets $G_{n,F_1} = B_{n,F_1} - \bigcup\{\overline{B}_{k,F_2} \mid k \leq n\}$ and $G_{n,F_2} = B_{n,F_2} - \bigcup\{\overline{B}_{k,F_1} \mid k \leq n\}$ are open. Moreover, G_{n,F_1} contains each $x \in F_1$ for which $n(x) = n$, and G_{n,F_2} contains each $y \in F_2$ for which $n(y) = n$. Let $G_{F_1} = \bigcup\{G_{n,F_1} \mid n \in I^+\}$ and $G_{F_2} = \bigcup\{G_{n,F_2} \mid n \in I^+\}$. Then $G_{F_1}, G_{F_2} \in \tau$, $G_{F_1} \cap G_{F_2} = \varnothing$, and $F_1 \subset G_{F_1}$ and $F_2 \subset G_{F_2}$.

THEOREM 5-2 (Nagata–Smirnov). $\langle S, \tau \rangle$ *is metrizable iff it is a* T_3*-space with a* σ*-locally finite base.*

Proof

Necessity. For each $n \in I^+$, the collection of open spheres of radius $1/n$ about each point of S is an open cover $\mathscr{B}_n$ that has an open locally finite refinement $\mathscr{B}'_n$ by paracompactness. The collection $\bigcup\{\mathscr{B}'_n \mid n \in I^+\}$ is clearly a σ-locally finite base for τ, which shows that the condition is necessary.

Sufficiency. Suppose that $\langle S, \tau \rangle$ is a T_3-space and has a σ-locally finite base $\mathscr{B} = \{B_{n,\lambda} \mid \lambda \in \Lambda_n, n \in I^+\}$. Let $\Lambda = \{\langle n, \lambda \rangle \mid n \in I^+, \lambda \in \lambda_n\}$ and suppose that the cardinality of Λ is α. $\langle S, \tau \rangle$ is normal by Lemma 5-2. By Lemma 5-1, each open subset is an F_σ. Thus, by Exercise 5-2, for each $\langle n, \lambda \rangle \in \Lambda$, there is a continuous function $f_{n,\lambda} : \langle S, \tau \rangle \to I^1$ such that $f_{n,\lambda}(x) > 0$ iff

$x \in B_{n,\lambda}$. For each $n \in I^+$, $\{B_{n,\lambda} \mid \lambda \in \Lambda_n\}$ is locally finite. Thus, for each $x \in S$, $f_{n,\lambda}(x) \neq 0$ for only finitely many values of λ. Hence

$$1 + \sum_{\lambda \in \Lambda_n} f_{n,\lambda}^2(x) \geq 1$$

and defines a continuous function on S. Let $g_{n,\lambda}: \langle S, \tau\rangle \to I^1$ be given by

$$g_{n,\lambda}(x) = f_{n,\lambda}(x)\left[1 + \sum_{\beta \in \Lambda_n} f_{n,\beta}^2(x)\right]^{-1/2}.$$

Then $g_{n,\lambda}(x) > 0$ iff $x \in B_{n,\lambda}$. For fixed $n \in I^+$ and fixed $x \in S$, $g_{n,\lambda}(x) \neq 0$ for only finitely many values of λ. Moreover,

$$\sum_{\lambda \in \Lambda_n} g_{n,\lambda}^2(x) < 1 \qquad \text{and} \qquad \sum_{\lambda \in \Lambda_n} [g_{n,\lambda}(x) - g_{n,\lambda}(y)]^2 < 2, \ \forall x, y \in S.$$

Let $h_{n,\lambda}(x) = 2^{-n/2} g_{n,\lambda}(x)$. Then

$$\sum_{\langle n,\lambda\rangle \in \Lambda} h_{n,\lambda}^2(x) = \sum_{n \in I^+} 2^{-n} \sum_{\lambda \in \Lambda_n} g_{n,\lambda}^2(x) < \sum_{n \in I^+} 2^{-n} = 1.$$

For each $x \in S$, we have a point $f(x) = h_{n,\lambda}(x) \in H^\alpha$. We show that f is a homeomorphism onto $f(S)$.

If $x, y \in S$ with $x \neq y$, then $\exists B_{n,\lambda} \in \mathscr{B}$ such that $x \in B_{n,\lambda}$, $y \in S - B_{n,\lambda}$, since $\langle S, \tau\rangle$ is a T_1-space. Then $h_{n,\lambda}(x) > 0$ and $h_{n,\lambda}(y) = 0$, which implies that $f(x) \neq f(y)$. Thus f is injective.

Let $x_0 \in S$ and $\varepsilon > 0$. Then $\exists N \in I^+$ such that $2^{-N} < \varepsilon^2/4$. By local finiteness, $\exists G \in \tau$ such that $x_0 \in G$ and G intersects only finitely many of the sets $B_{n,\lambda}$ with $n \leq N$. Denote these by B_{n_i,λ_i} ($n_i \leq N$) for $i = 1, \ldots, K$. Since each function $h_{n,\lambda}$ is continuous, for each $\langle n, \lambda\rangle \in \Lambda$, $\exists B_{n,\lambda} \in \mathscr{B}$ such that $x_0 \in B_{n,\lambda}$ and

$$|h_{n,\lambda}(x_0) - h_{n,\lambda}(x)| < \frac{\varepsilon}{\sqrt{2K}} \qquad \forall x \in B_{n,\lambda}.$$

Let $\hat{G} = G \cap \bigcap_{i=1}^K B_{n_i,\lambda_i}$. Then $\hat{G} \in \tau$ and contains x_0. If $\langle n, \lambda\rangle \in \Lambda - \{\langle n_i, \lambda_i\rangle \mid i = 1, \ldots, K\}$, then $h_{n,\lambda}(x_0) = h_{n,\lambda}(x) = 0$, $\forall x \in \hat{G}$. Thus we have for $x \in \hat{G}$,

$$\sum_{n \leq N, \lambda \in \Lambda_n} [h_{n,\lambda}(x_0) - h_{n,\lambda}(x)]^2 < K\left(\frac{\varepsilon}{\sqrt{2K}}\right)^2 = \frac{\varepsilon^2}{2}$$

and

$$\sum_{n > N, \lambda \in \Lambda_n} [h_{n,\lambda}(x_0) - h_{n,\lambda}(x)]^2 = \sum_{n > N} 2^{-n} \sum_{\lambda \in \Lambda_n} [g_{n,\lambda}(x_0) - g_{n,\lambda}(x)]^2$$
$$\leq 2 \sum_{n > N} 2^{-n} = 2(2^{-N}) < 2\left(\frac{\varepsilon^2}{4}\right) = \frac{\varepsilon^2}{2}.$$

Thus

$$d^{\alpha}(f(x_0), f(x)) = \left[\sum_{\langle n, \lambda \rangle \in \Lambda} [h_{n,\lambda}(x_0) - h_{n,\lambda}(x)]^2 \right]^{1/2} < \varepsilon \qquad \forall x \in \hat{G}.$$

This implies that f is continuous at x_0 and hence on S, since x_0 was arbitrary.

It remains to show that f is open. Let $G \in \tau$, and let $x \in G$. Then $\exists B_{n,\lambda} \in \mathscr{B}$ such that $x \in B_{n,\lambda} \subset G$. Then $h_{n,\lambda}(x) = \delta > 0$. If $f(y) \in f(S)$ such that $d^{\alpha}(f(x), f(y)) < \delta$, then $h_{n,\lambda}(y) > 0$ and $y \in B_{n,\lambda}$. Hence $f^{-1}(S_{d^{\alpha}}(f(x); \delta)) \subset G$, which implies that $f(G)$ is open.

Using a method analogous to the proof of Stone's theorem that every metric space is paracompact, one can show that every open cover of a metric space has an open σ-discrete refinement. Thus a metric space is a T_3-space with a σ-discrete base. Also, every discrete collection is locally finite. These remarks establish the following metrization theorem of R. H. Bing.

THEOREM 5-3 (Bing). *$\langle S, \tau \rangle$ is metrizable iff it is a T_3-space with a σ-discrete base.*

EXERCISES

5-1. Let $d(x, y) = 0$ if $x = y$ and $d(x, y) = 1$ if $x \neq y$, where $x, y \in S$ and S has the discrete topology 2^S. Show that d metrizes $\langle S, 2^S \rangle$; i.e., show that τ_d is the discrete topology 2^S.

5-2. Show that if $\langle S, \tau \rangle$ is normal and $G \in \tau$ is an F_σ set, there is a continuous function $f: \langle S, \tau \rangle \to I^1$ such that $f(x) > 0$ iff $x \in G$, and thus $f^{-1}\{0\} = S - G$.

5-3. Prove that a compact Hausdorff space is metrizable iff it is second countable. (This is Urysohn's second metrization theorem.)

5-4. Show that Urysohn's metrization theorems are corollaries to Bing's metrization theorem.

5-5. $\langle S, \tau \rangle$ is *locally metrizable* iff each point of S is contained in an open metrizable subspace of $\langle S, \tau \rangle$. Show that if $\langle S, \tau \rangle$ is a normal space which is the union of a locally finite family of metrizable subspaces, then $\langle S, \tau \rangle$ is metrizable.

5-6. Show that a locally metrizable Hausdorff space is metrizable iff it is paracompact. (This result is due to Smirnov.)

5-7. Show that the product of two metrizable spaces is metrizable.

★5-2 Semimetrizable and a-Metrizable Spaces

In this section we investigate two generalizations of metrizability, semimetrizability and a-metrizability. We present a characterization of semimetrizable spaces in terms of "indexed" neighborhoods. Also, we show that the classes of semimetrizable and a-metrizable spaces are identical for the class of T_1-spaces.

DEFINITION 5-4. A topological space $\langle S, \tau \rangle$ is *semimetrizable* iff there is a semimetric ρ on S such that for each $x \in S$, the collection $\{S_\rho(x; r) \mid r > 0\}$ is a local base at x for τ; i.e., the semimetric topology τ_ρ induced on S by ρ is identically τ. Such a semimetric ρ is said to be *admissible* on $\langle S, \tau \rangle$.

Every developable T_1-space is semimetrizable. Details are left to the reader as an exercise. For our semimetrization theorem, we need the following lemma on indexed neighborhood systems for a topological space $\langle S, \tau \rangle$. Let $\{N_i \mid i \in I^+\}$ be a family of functions, indexed by the set I^+ of positive integers, which assign to each $x \in S$ a subset $N_i(x)$ of S and consider the following properties:

1. For each $x \in S$, $\{N_i(x) \mid i \in I^+\}$ is a local base at x.
2. For each $i \in I^+$, $x \in N_i(y)$ implies that $y \in N_i(x)$.
3. For $m, n \in I^+$, $\exists k \in I^+$ such that $\forall x \in S$, $N_k(x) \subset N_m(x) \cap N_n(x)$.

3′. For each $x \in S$ and $m, n \in I^+$, $N_m(x) \subset N_n(x)$ if $m > n$.

LEMMA 5-3. *Condition* 3′ *implies* 3, *and if there exists a family* $\{N_i \mid i \in I^+\}$ *satisfying* 1–3, *then there is a family* $\{N_i' \mid i \in I^+\}$ *satisfying* 1, 2, *and* 3′.

Proof. Choosing k to be the larger of m and n suffices for the first part. The second part is proved by setting $N_1' = N_1$ and $N_{i+1}' = N_k$ (for $i \in I^+$), where k is the first positive integer such that $N_k(x) \subset N_i'(x) \cap N_{i+1}(x)$ for all $x \in S$.

THEOREM 5-4. *A necessary and sufficient condition that a* T_1-*space* $\langle S, \tau \rangle$ *be semimetrizable is that there exists a family* $\{N_i \mid i \in I^+\}$ *of functions which assign to each* $x \in S$ *a subset* $N_i(x)$ *of* S *such that conditions* 1–3 *are satisfied.*

Proof

Necessity. Let ρ be any admissible semimetric on $\langle S, \tau\rangle$, and let $N_i(x) = S_\rho(x; 1/i)$, $\forall i \in I^+$ and $\forall x \in S$. Then $\{N_i \mid i \in I^+\}$ satisfies 1, 2, and 3′. By Lemma 5-3, $\{N_i \mid i \in I^+\}$ satisfies condition 3 also.

Sufficiency. Let $\{N_i \mid i \in I^+\}$ be any family of functions satisfying 1–3. By Lemma 5-3, there is a family $\{N_i' \mid i \in I^+\}$ satisfying 1, 2, and 3′. Then $\bigcap\{N_i'(x) \mid i \in I^+\} = \{x\}$. Define a semimetric ρ on S as follows:

$$\rho(x, y) = \begin{cases} 0 & \text{if } x = y \\ \sup\left\{\dfrac{1}{i} \,\middle|\, y \notin N_i'(x),\ i \in I^+\right\} & \text{if } x \neq y. \end{cases}$$

For each $x \in S$ and each $i \in I^+$, $S_\rho(x; 1/i) = N_i'(x)$, and thus $\langle S, \tau\rangle$ is semimetrizable, since $\{N_i'(x) \mid i \in I^+\}$ is a local base at x.

It is possible that the existence of a semimetric with a countable distance set, such as that which we have just constructed, could be useful in the construction of induction proofs in a semimetrizable space. With the introduction of such a semimetric on a nontotally disconnected space, however, we lose all manner of continuity of the semimetric. We motivate our definition of a-metrizability by means of the following example.

EXAMPLE 5-1. Let S be the closed upper half-plane in E^2, and let L denote the x-axis. For each $p \in S$ and each $r > 0$, define a neighborhood $N_r(p)$ as follows:

(1) $N_r(p) = S_d(p; r) \cap S$ if $p \in S - L$.

(2) $N_r(p) = [S_d(p; r) \cap (S - L)] \cup \{p\}$ if $p \in L$.

Here d is the usual metric for E^2 and $S_d(p; r) = \{x \in E^2 \mid d(p, x) < r\}$. Then a topology τ is generated for S by the collection $\{N_r(p) \mid p \in S, r > 0\}$. The space $\langle S, \tau\rangle$ is Hausdorff but not regular; first countable but not second countable; separable but not hereditarily separable. Define a metric ρ on S as follows:

$$\rho(x, y) = \begin{cases} 0 & \text{if } x = y \\ d(x, y) + \frac{1}{2} & \text{if } x \notin L \text{ or } y \notin L \\ d(x, y) + 1 & \text{if } x, y \in L. \end{cases}$$

For each $p \in S$, we have that $S_\rho(p; r) = N_{r-(1/2)}(p)$, $\frac{1}{2} < r \leq 1$. Thus, although ρ does not metrize $\langle S, \tau\rangle$, we observe that for each $p \in S$ the collection $\{S_\rho(p; r) \mid r > \frac{1}{2}\}$ is a local base at p for τ. Abstracting from this example, we obtain a natural formulation of "a-metrizability."

DEFINITION 5-5. Let $a \geq 0$. $\langle S, \tau \rangle$ is *a-metrizable* iff there exists a metric ρ on S such that $\forall p \in S$, the collection $\{S_\rho(p; r) \mid r > a\}$ is a local base at p for τ. The metric ρ is called an *admissible a-metric* for $\langle S, \tau \rangle$.

In view of Definition 5-1 and the remarks which follow it, we see that the class of metrizable spaces is precisely the class of 0-metrizable spaces. One can easily show that a-metrizability is a hereditary topological property. The theorem and example which follow establish that a-metrizability is a generalization of semimetrizability for topological spaces possessing no T_i-separation. Indeed, the class of a-metrizable spaces occupies a position strictly intermediate to the classes of semimetrizable and first countable spaces.

THEOREM 5-5. *A necessary and sufficient condition that* $\langle S, \tau \rangle$ *be semimetrizable is that it be an a-metrizable* T_0*-space.*

Proof

Sufficiency. Let ρ be an admissible a-metric for $\langle S, \tau \rangle$. For $x, y \in S$, let (1) $d(x, y) = 0$ iff $x = y$ and (2) $d(x, y) = \rho(x, y) - a$ if $x \neq y$. Since $\langle S, \tau \rangle$ is a T_0-space, we have $\rho(x, y) > a$ and hence $d(x, y) > 0$ if $x \neq y$. Since $S_d(p; r - a) = S_\rho(p; r)$ for $r > a$ and each $p \in S$, d is an admissible semimetric on $\langle S, \tau \rangle$.

Necessity. Let d be a bounded, admissible semimetric on $\langle S, \tau \rangle$, and set $a = \sup \{d(x, y) - d(x, z) - d(z, y) \mid x, y, z \in S\}$. For $x, y \in S$, let (1) $\rho(x, y) = 0$ iff $x = y$ and (2) $\rho(x, y) = d(x, y) + a$ if $x \neq y$. Then ρ is a metric on S. Since $S_\rho(p; r + a) = S_d(p; r)$ for $r > 0$ and each $p \in S$, ρ is an admissible a-metric for $\langle S, \tau \rangle$.

EXAMPLE 5-2. Let $S = \{b, c, d\}$ and $\tau = \{\varnothing, \{b\}, \{c, d\}, S\}$. $\langle S, \tau \rangle$ is not semimetrizable, since it is not a T_0-space. Let $\rho(c, d) = \rho(d, c) = 1$, $\rho(b, c) = \rho(c, b) = \rho(b, d) = \rho(d, b) = 2$, and $\rho(x, y) = 0$ if $x, y \in S$ with $x = y$. If $1 < r \leq 2$, then $S_\rho(b; r) = \{b\}$ and $S_\rho(c; r) = S_\rho(d; r) = \{c, d\}$. Hence $\langle S, \tau \rangle$ is 1-metrizable.

In constructing an a-metric ρ from the semimetric d in the proof of Theorem 5-5, we set

$$a = \sup \{d(x, y) - d(x, z) - d(z, y) \mid x, y, z \in S\},$$

but this choice of a is not unique. Indeed, if ρ is an admissible a-metric for $\langle S, \tau \rangle$ and $b > 0$, one can set $\rho'(x, y) = \rho(x, y) + b$ if $x \neq y$ [$\rho'(x, x) = 0$, $\forall x \in S$], and ρ' will be an admissible $(a + b)$-metric for $\langle S, \tau \rangle$. That is, although ρ' clearly induces the discrete topology on S, the collection of all

ρ'-spheres of radius exceeding $a + b$ constitutes a base for τ. At the other end of the spectrum, the number a cannot be reduced below the value used in the proof of Theorem 5-5, but of course one can diminish the "scale," e.g., by defining

$$\rho_k(x, y) = \frac{\rho'(x, y)}{k(\rho'(x, y) + 1)}$$

for any $k \in I^+$. This gives an admissible a_k-metric for $\langle S, \tau \rangle$ with

$$a_k = \frac{a + b}{k(a + b + 1)}$$

as small as one may wish, and the associated semimetric has correspondingly small deficiency in the triangle inequality. This is no surprise, however, since all distances are then small. Observe that the (pointwise) limit function $\rho = \lim_{k \to \infty} \rho_k$ is a pseudometric inducing the indiscrete topology on S.

EXERCISES

5-8. Show that a developable T_1-space is semimetrizable.

5-9. Prove that the space $\langle S, \tau \rangle$ described in Example 5-1 has the topological properties that are listed.

5-10. Show that if $\langle S, \tau \rangle$ is an a-metrizable T_0-space, then it is a T_1-space.

5-11. Prove that if $\langle S, \tau \rangle$ is a-metrizable, then it is first countable.

5-12. Show that each a-metrizable space has an admissible bounded a-metric.

★5-3 Uniform Spaces

In Section 2-2 it was shown that every completely regular space is uniformizable. Also, it was remarked that every uniform space is completely regular. In Section 1-8 it was stated that the class of uniformizable topological spaces is the same as the class of proximity spaces and includes the class of metric spaces as a special subclass. Thus it is natural to inquire under what conditions a uniform space is metrizable. This question is answered in the discussion that follows.

DEFINITION 5-6. A uniform space $\langle S, \mathscr{U} \rangle$ is *pseudometrizable* iff there is a pseudometric d on S such that $\tau_d = \tau_{\mathscr{U}}$.

THEOREM 5-6. *A uniform space $\langle S, \mathcal{U}\rangle$ is pseudometrizable iff $\mathcal{U}$ has a countable base.*

Proof. If ρ is a pseudometric admissible on $\langle S, \mathcal{U}\rangle$, then $\mathcal{B} = \{W_n \mid n \in I^+\}$, where $W_n = \{\langle x, y\rangle \in S \times S \mid \rho(x, y) < 1/n\}$ is a countable base for $\mathcal{U}$. Conversely, suppose that $\langle S, \mathcal{U}\rangle$ is a uniform space and $\mathcal{B} = \{V_n \mid n \in I^+\}$ is a countable base for $\mathcal{U}$. Using axioms (U4) and (U5) of a uniformity, we can define inductively a base $\mathcal{B}' = \{U_n \mid n \in I^+\}$ for $\mathcal{U}$ such that $U_n^{-1} = U_n$, $U_n \subset V_n$, and $U_n \circ U_n \circ U_n \subset U_{n-1}, \forall n \in I^+$. Define a mapping $\varphi: S \times S \to R$ by $\varphi(x, y) = \inf\{2^{-n} \mid \langle x, y\rangle \in U_n, n \in I^+\}$. Let $\rho(x, y) = \inf\{\sum_{i=1}^{k} \varphi(x_{i-1}, x_i) \mid \langle x_0, \ldots, x_k\rangle$ is any finite sequence of points of S such that $x_0 = x$ and $x_k = y\}$. The reader can easily show that ρ is a pseudometric on S. It remains to show that $\tau_\rho = \tau_{\mathcal{U}}$. Let $\varepsilon > 0$ and choose $N \in I^+$ such that $2^{-N} < \varepsilon$. If $\langle x, y\rangle \in U_N$, then $\rho(x, y) \leq \varphi(x, y) \leq 2^{-N} < \varepsilon$, which implies that $\langle x, y\rangle \in W_\varepsilon$. Hence $U_N \subset W_\varepsilon$ and $\tau_\rho \subset \tau_{\mathcal{U}}$. Clearly, $\varphi(x_0, x_1) \leq 2\varphi(x_0, x_1)$. Suppose that $\varphi(x_0, x_k) \leq 2\sum_{i=1}^{k} \varphi(x_{i-1}, x_i)$ for all positive integers less than k. Let $S = \sum_{i=1}^{k} \varphi(x_{i-1}, x_i)$ and let j be the largest integer for which $\sum_{i=1}^{j} \varphi(x_{i-1}, x_i) \leq S/2$. Then $\sum_{i=j+1}^{k} \varphi(x_{i-1}, x_i) \geq S/2$ and $\varphi(x_j, x_{j+1}) \leq S$. By the induction hypothesis, we have $\varphi(x_0, x_j) \leq 2(S/2) = S$. Also, $\varphi(x_{j+1}, x_k) \leq S$. Let n be the smallest positive integer such that $2^{-n} \leq 2^S$. Then $\langle x_0, x_j\rangle, \langle x_j, x_{j+1}\rangle$, and $\langle x_{j+1}, x_k\rangle$ are all in U_{n+1}, which implies that $\langle x_0, x_k\rangle \in U_n$. Thus $\varphi(x_0, x_k) \leq 2^{-n} \leq 2S = 2\sum_{i=1}^{k} \varphi(x_{i-1}, x_i)$. This establishes the latter inequality for all $k \in I^+$. Suppose that $\langle x, y\rangle \in W_{2^{-(n+1)}}$ for fixed $n \in I^+$. Then $\rho(x, y) < 2^{-(n+1)}$, which implies that $\sum_{i=1}^{k} \varphi(x_{i-1}, x_i) < 2^{-(n+1)}$ for some finite sequence $\langle x_0, \ldots, x_k\rangle$ such that $x_0 = x$ and $x_k = y$. Then $\varphi(x_0, x_k) \leq 2\sum_{i=1}^{k} \varphi(x_{i-1}, x_i) < 2 \cdot 2^{-(n+1)} = 2^{-n}$. Hence $\langle x, y\rangle \in U_n$, which implies that $W_{2^{-(n+1)}} \subset U_n$ and $\tau_{\mathcal{U}} \subset \tau_\rho$. Thus $\tau_\rho = \tau_{\mathcal{U}}$.

COROLLARY. *A Hausdorff uniform space $\langle S, \mathcal{U}\rangle$ is metrizable iff $\mathcal{U}$ has a countable base.*

Proof. The proof follows immediately from Theorem 5-6, since the pseudometric ρ is actually a metric if the space $\langle S, \mathcal{U}\rangle$ is Hausdorff.

A mapping f of a uniform space $\langle S, \mathcal{U}\rangle$ into a uniform space $\langle T, \mathcal{V}\rangle$ is *uniformly continuous* iff $\forall V \in \mathcal{V}\ \exists U \in \mathcal{U}$ such that $\langle f(x), f(y)\rangle \in V$ whenever $\langle x, y\rangle \in U$. If f is a bijection with f and f^{-1} both uniformly continuous, then f is a *uniform isomorphism*. Properties of uniform spaces that are preserved by uniform isomorphisms are called *uniform invariants.*

If $\langle S_\alpha, \mathcal{U}_\alpha\rangle$ is a uniform space $\forall \alpha \in \Lambda$, then $\prod_\Lambda \langle S_\alpha, \mathcal{U}_\alpha\rangle$ is a uniform space in which one assigns to $\prod_\Lambda S_\alpha$ the *product uniformity*, which is the smallest uniformity for which π_α is uniformly continuous $\forall \alpha \in \Lambda$. The reader should verify that the topology induced by the product uniformity on $\prod_\Lambda S_\alpha$ is precisely the Tychonoff product topology $\prod_\Lambda \tau_{\mathscr{U}_\alpha}$.

If ρ is a pseudometric on a uniform space $\langle S, \mathscr{U} \rangle$, the *pseudometric uniformity* induced by ρ is the smallest uniformity that makes ρ uniformly continuous on $\langle S \times S, \mathscr{U} \times \mathscr{U} \rangle$. It has as a subbase the collection of all sets $W_\varepsilon = \{\langle x, y \rangle \in S \times S \mid \rho(x, y) < \varepsilon\}$, $\forall \varepsilon > 0$.

Our final result in this section states that the uniformity $\mathscr{U}$ of a uniform space $\langle S, \mathscr{U} \rangle$ is generated by the family $\{\rho_\alpha \mid \alpha \in \Lambda\}$ of *all* pseudometrics that are uniformly continuous on $\langle S \times S, \mathscr{U} \times \mathscr{U} \rangle$. Thus it must have as a subbase the collection of all sets $W_{\varepsilon,\alpha} = \{\langle x, y \rangle \in S \times S \mid \rho_\alpha(x, y) < \varepsilon\}$, $\forall \varepsilon > 0$ and $\forall \alpha \in \Lambda$. This family of pseudometrics is also called a family of *gauges* for $\langle S, \mathscr{U} \rangle$.

LEMMA 5-4. *If ρ is a pseudometric on $\langle S, \mathscr{U} \rangle$, then ρ is uniformly continuous on $S \times S$ iff $W_\varepsilon = \{\langle x, y \rangle \in S \times S \mid \rho(x, y) < \varepsilon\} \in \mathscr{U}$, $\forall \varepsilon > 0$.*

Proof. See Pervin, *Foundations of General Topology*, pp. 186–87.

THEOREM 5-7. *If $\langle S, \mathscr{U} \rangle$ is a uniform space, then the uniformity $\mathscr{U}$ is generated by the family of all pseudometrics that are uniformly continuous on $S \times S$.*

Proof. Let $\{\rho_\alpha \mid \alpha \in \Lambda\}$ be the family of all pseudometrics on $\langle S, \mathscr{U} \rangle$ such that ρ_α is uniformly continuous on $S \times S$ $\forall \alpha \in \Lambda$. For each $\alpha \in \Lambda$ and $\varepsilon > 0$, let $W_{\varepsilon,\alpha} = \{\langle x, y \rangle \in S \times S \mid \rho_\alpha(x, y) < \varepsilon\}$. Then $W_{\varepsilon,\alpha} \in \mathscr{U}$, $\forall \alpha \in \Lambda$ and $\forall \varepsilon > 0$, by Lemma 5-4. Now if $U \in \mathscr{U}$, there is a pseudometric ρ_α on $\langle S, \mathscr{U} \rangle$ such that $W_{\varepsilon,\alpha} \subset U$ for some $\varepsilon > 0$ by Theorem 5-6. Since ρ_α is uniformly continuous in the uniformity it generates, and thus in any larger uniformity (such as $\mathscr{U}$), the uniformity generated by $\{\rho_\alpha \mid \alpha \in \Lambda\}$ contains $\mathscr{U}$.

EXERCISES

5-13. Prove that every topological space $\langle S, \tau \rangle$ is quasiuniformizable. (*Hint:* For each $G \in \tau$, let $S_G = (G \times G) \cup [(S - G) \times S]$. Then show that $\mathscr{S} = \{S_G \mid G \in \tau\}$ is a subbase for a quasiuniformity which induces the topology τ on S.)

5-14. Show that the mapping ρ defined in the proof of Theorem 5-6 is a pseudometric.

5-15. Show that the uniformity generated by a family $\mathscr{P} = \{\rho_\alpha \mid \alpha \in \Lambda\}$ of pseudometrics on S is the smallest such that the identity mapping of S into $\langle S, \rho_\alpha \rangle$ is uniformly continuous $\forall \alpha \in \Lambda$.

★5-4 Topological Groups

In this section we introduce the concept of a "topological group." We show that a topological group is uniformizable (hence completely regular) by defining a uniformity for the topological group that generates the same topology. Then we show that this uniformity is generated by the family of all invariant pseudometrics on the topological group which are uniformly continuous.

DEFINITION 5-7. A triple $\langle G, \cdot, \tau \rangle$ is a *topological group* iff $\langle G, \cdot \rangle$ is a group, $\langle G, \tau \rangle$ is a topological space, and the function $h: G \times G \to G$ given by $h(x, y) = x \cdot y^{-1}$ is continuous relative to the product topology on $G \times G$.

EXAMPLE 5-3. Every group with the discrete or indiscrete topology is a topological group.

EXAMPLE 5-4. Euclidean n-space with the usual topology and the operation of vector addition is a topological group.

EXAMPLE 5-5. The unit circle in the complex plane with the relative topology and the operation of complex multiplication is a topological group.

DEFINITION 5-8. Let $\langle G, \cdot, \tau \rangle$ be a *topological group* and $X, Y \in 2^G$. Then $X \cdot Y = \{z \in G \mid \exists x \in X, y \in Y \text{ with } x \cdot y = z\}$ and $Y^{-1} = \{x \in G \mid x^{-1} \in Y\}$.

DEFINITION 5-9. For each $a \in G$, let $L_a(x) = a \cdot x$ and $R_a(x) = x \cdot a$ where $\langle G, \cdot, \tau \rangle$ is a *topological group*. It is easily seen that L_a and R_a are homeomorphisms $\forall a \in G$.

DEFINITION 5-10. Let $\langle G, \cdot, \tau \rangle$ be a *topological group* and $\mathscr{V}_e$ be the neighborhood system of the identity e. For each $U \in \mathscr{V}_e$, let $U_L = \{x, y\rangle \mid x^{-1} \cdot y \in U\}$ and $U_R = \langle x, y\rangle \mid x \cdot y^{-1} \in U\}$. The collections $\{U_L \mid U \in \mathscr{V}_e\}$ and $\{U_R \mid U \in \mathscr{V}_e\}$ are, respectively, bases for the *left uniformity* $\mathscr{L}$ and the *right uniformity* $\mathscr{R}$ for G. Also, G has a *two-sided uniformity* $\mathscr{U}$, which has $\mathscr{L} \cup \mathscr{R}$ as a subbase. If $\langle G, \cdot, \tau \rangle$ is Abelian, then $\mathscr{R}$, $\mathscr{L}$, and $\mathscr{U}$ coincide.

THEOREM 5-8. *If* $\langle G, \cdot, \tau \rangle$ *is a topological group and the uniformities* $\mathscr{L}$, $\mathscr{R}$, *and* $\mathscr{U}$ *are as defined in Definition* 5-10, *then* τ *is the topology induced on* S *by each of them.*

Proof. Let $\mathscr{V}_e$ be the neighborhood system of e. We show that $\mathscr{B} = \{U_L \mid U \in \mathscr{V}_e\}$ is a base for the uniformity $\mathscr{L}$. An entirely similar argument holds for $\mathscr{R}$. Each $U_L \in \mathscr{B}$ contains the diagonal Δ of $G \times G$, since $x \cdot x^{-1} = e \in U$. If $U_L \in \mathscr{B}$, then $U_L^{-1} = \{\langle y, x\rangle \mid \langle x, y\rangle \in U_L\} = \{\langle y, x\rangle \mid y^{-1} \cdot x \in U^{-1}\} \in \mathscr{B}$, since $U^{-1} \in \mathscr{V}_e$. For $U_L \in \mathscr{B}$, $\exists V \in \mathscr{V}_e$ such that $V \cdot V \subset U$. If $\langle x, z\rangle \in V_L$ and $\langle z, y\rangle \in V_L$, then $x^{-1} \cdot z$ and $z^{-1} \cdot y$ are in V. Hence $x^{-1} \cdot y \in U$ and $\langle x, y\rangle \in U_L$. Consequently, $V_L \circ V_L \subset U_L$. Let U_L and U'_L be in $\mathscr{B}$. Then $U_L \cap U'_L \in \mathscr{B}$, for $U \cap U' \in \mathscr{V}_e$.

It remains to show that $\tau_{\mathscr{L}} = \tau$. Suppose that $U \in \tau$ and $x \in U$. Then $L_{x^{-1}}(U) = V$ is an open neighborhood of e. Moreover, $V_L[x] = \{y \mid x^{-1} \cdot y \in V\} = \{y \mid y \in xV\} = L_x(L_{x^{-1}}(U))$; i.e., $U \in \tau_{\mathscr{L}}$. Assume that $U \in \tau_{\mathscr{L}}$. For each $x \in U$, there is a $V_L[x] = L_x(V) \subset U$. Hence $U \in \tau$ and $\tau_{\mathscr{L}} = \tau$. Moreover, for each $U \in \mathscr{L} \cup \mathscr{R}$, we have (1) $\Delta \subset U$, (2) $U^{-1} \in \mathscr{L} \cup \mathscr{R}$, and (3) $\exists V \in \mathscr{L} \cup \mathscr{R}$ such that $V \circ V \subset U$, since the elements of $\mathscr{R}$ and $\mathscr{L}$ (separately) have these properties. Thus $\mathscr{L} \cup \mathscr{R}$ is a subbase for a uniformity $\mathscr{U}$, and it follows as before that $\tau = \tau_{\mathscr{U}}$.

DEFINITION 5-11. Let $\langle G, \cdot, \tau\rangle$ be a topological group.

(1) A pseudometric ρ on G is *left-invariant* (*right-invariant*) iff $\rho(x, y) = \rho(g \cdot x, g \cdot y)$ ($\rho(x, y) = \rho(x \cdot g, y \cdot g)$) for all $\langle x, y\rangle \in G \times G$ and all $g \in G$. We say that ρ is *invariant* iff ρ is both left- and right-invariant.

(2) Let $\mathscr{V}_x$ be the neighborhood system of $x \in G$ and $V \in \mathscr{V}_x$. V is *invariant* iff $g^{-1} \cdot V \cdot g = V$ for all $g \in G$.

THEOREM 5-9. *The uniformity $\mathscr{L}$ (respectively, $\mathscr{R}$) is generated by the family of all left-invariant (respectively, right-invariant) pseudometrics that are uniformly continuous on $G \times G$.*

Proof. Let $\mathscr{P}$ be the family of all left-invariant pseudometrics that are uniformly continuous on $G \times G$, and let $\rho \in \mathscr{P}$. Then $V_{\rho,\alpha} = \{z \in G \mid z = x^{-1} \cdot y,\ x, y \in G$ and $\rho(x, y) = \rho(e, x^{-1} \cdot y) < \alpha\} \in \mathscr{V}_e$. Hence $(V_{\rho,\alpha})_L = \{\langle x, y\rangle \mid x^{-1} \cdot y \in V_{\rho,\alpha}\}$, and $\{(V_{\rho,\alpha})_L \mid \rho \in \mathscr{P}, \alpha > 0\}$ generates $\mathscr{L}$. A similar argument holds in the case of $\mathscr{R}$.

THEOREM 5-10. *Let $\mathscr{I}$ be the family of all neighborhoods of the identity e that are invariant. Then $\mathscr{I}$ is a base for the neighborhood system of e iff the family of all invariant pseudometrics that are uniformly continuous on $G \times G$ generates a uniformity whose induced topology is τ.*

Proof. Let $\mathscr{P}$ be a family of invariant pseudometrics that are uniformly continuous on $G \times G$ and such that $\mathscr{P}$ generates τ. Then $\{y \mid \rho(e, y) < \alpha,\ \alpha > 0\} = V_{\rho,\alpha} \in \tau$, $\forall \rho \in \mathscr{P}$. If $x \in V_\alpha$, then $g^{-1} \cdot x \cdot g \in V_{\rho,\alpha}$ for each $g \in G$, since $\rho(e, x) = \rho(e, g^{-1} \cdot x \cdot g)$. Thus $V_{\rho,\alpha} \in \mathscr{I}$. Conversely, suppose that U

is an invariant neighborhood of the identity e. Then $g^{-1} \cdot U \cdot g = U, \forall g \in G$, and so $g \cdot U = U \cdot g$. This implies that $U_L = U_R$, and both are in $\mathscr{L} \cup \mathscr{R}$, which is generated by the family of all invariant pseudometrics that are uniformly continuous on $G \times G$ by Theorem 5-9. By Theorem 5-8, $\mathscr{L} \cup \mathscr{R}$ generates a uniformity $\mathscr{U}$ such that $\tau_{\mathscr{U}} = \tau$.

EXERCISE

5-16. Show that the mappings L_a and R_a defined in Definition 5-9 are homeomorphisms.

CHAPTER 6
Convergence and Completeness

6-1 Sequences and Filters

In this chapter we discuss the convergence of sequences in metric spaces and the convergence of filters in uniform spaces. The notion of "completeness" is defined and characterized. Also, we describe a method for topologically embedding any metric space densely in a complete metric space. As an application, we prove the famous Contraction Mapping Theorem of Banach.

Recall that a sequence $\{x_n\}_{n \in I^+}$ in a metric space $\langle M, d \rangle$ is a function $x : I^+ \to M$ such that $x(n) = x_n$, $\forall n \in I^+$. Recall also that $\{x_n\}_{n \in I^+}$ converges to $x_0 \in M$ (in symbols, $x_n \to x_0$) iff $\lim d(x_n, x_0) = 0$; i.e., $\forall \varepsilon > 0$, $\exists N \in I^+$ such that $d(x_n, x_0) < \varepsilon$, $\forall n \geq N$. Convergent sequences in a metric space have the following important property, which was discovered by the French mathematician Cauchy.

DEFINITION 6-1. A sequence $\{x_n\}_{n \in I^+}$ in a metric space $\langle M, d \rangle$ is a *Cauchy sequence* iff $\forall \varepsilon > 0$, $\exists N \in I^+$ such that $d(x_m, x_n) < \varepsilon$, $\forall m, n \geq N$.

THEOREM 6-1. *Every convergent sequence in a metric space $\langle M, d \rangle$ is a Cauchy sequence.*

Proof. Let $\{x_n\}_{n \in I^+}$ be any sequence in $\langle M, d \rangle$, and suppose that $x_n \to x_0 \in M$. Let $\varepsilon > 0$. Then $\exists N \in I^+$ such that $d(x_k, x_0) < \varepsilon/2$, $\forall k \geq N$. Then $d(x_m, x_n) \leq d(x_m, x_0) + d(x_0, x_n) < \varepsilon$, $\forall m, n \geq N$. Hence $\{x_n\}_{n \in I^+}$ is a Cauchy sequence.

The converse of Theorem 6-1 is not true, as the following example shows; i.e., a Cauchy sequence in an arbitrary metric space need not converge.

EXAMPLE 6-1. Let $M = \{\text{rationals}\}$ and $d(x, y) = |x - y|$ for $x, y \in M$. Then $\langle M, d \rangle$ is the subspace of E^1 consisting of the rationals with the Euclidean subspace topology. Let $x_n = (1 + 1/n)^n$, $\forall n \in I^+$. Then $\{x_n\}_{n \in I^+}$ is a Cauchy sequence in E^1 (hence in $\langle M, d \rangle$), since it converges to $e \in R$. However, $\{x_n\}_{n \in I^+}$ does not converge in $\langle M, d \rangle$, since e is not rational.

For spaces that are not first countable, sequences are inadequate to discuss convergence, since in such a space $\langle S, \tau \rangle$ a sequence $x : I^+ \to S$ may have

$x_0 \in S$ as a limit point of $x(I^+) = \{x_n \mid n \in I^+\}$ but not have any subsequence converging to x_0. To avoid this difficulty in such spaces, and in uniform spaces in particular, we discuss convergence in terms of filters.

DEFINITION 6-2. Let $\langle S, \mathscr{U}\rangle$ be a uniform space. A *filterbase* in $\langle S, \mathscr{U}\rangle$ is a family $\mathscr{B} = \{A_\alpha \mid a \in \Lambda\}$ of nonempty subsets of S such that $\forall \alpha, \beta \in \Lambda\ \exists \gamma \in \Lambda$ such that $A_\gamma \subset A_\alpha \cap A_\beta$. If $\mathscr{B}$ is a filterbase in $\langle S, \mathscr{U}\rangle$ and $\mathscr{F}(\mathscr{B}) = \mathscr{B} \cup \{A \subset S \mid A \supset A_\alpha$ for some $\alpha \in \Lambda\}$, then $\mathscr{F}(\mathscr{B})$ is the *filter* generated by $\mathscr{B}$.

The reader should verify (as an easy exercise) that a filter in $\langle S, \mathscr{U}\rangle$ is a collection $\mathscr{F}$ of nonempty subsets of S, which is closed under finite intersections and contains every superset of every member of the collection. We introduce next the concepts "limit of a filter," "adherent point of a filter," "Cauchy filter," and "ultrafilter."

DEFINITION 6-3. Let $\mathscr{F}$ be a filter in a uniform space $\langle S, \mathscr{U}\rangle$. Then $\mathscr{F}$ *converges* to $x \in S$ iff $U[x] \in \mathscr{F}$, $\forall U \in \mathscr{U}$. We call x a *limit point* of the filter $\mathscr{F}$. Moreover, a point $x \in S$ is an *adherent point* of the filter $\mathscr{F}$ iff $x \in \bar{F}$, $\forall F \in \mathscr{F}$, and we say that the filter $\mathscr{F}$ *accumulates* at x in that case.

DEFINITION 6-4. A filter $\mathscr{F}$ in a uniform space $\langle S, \mathscr{U}\rangle$ is a *Cauchy filter* iff $\forall U \in \mathscr{U}\ \exists F \in \mathscr{F}$ such that $F \times F \subset U$.

DEFINITION 6-5. An *ultrafilter* in $\langle S, \mathscr{U}\rangle$ is a filter $\mathscr{F}$ in $\langle S, \mathscr{U}\rangle$ that is maximal in the collection of all filters in $\langle S, \mathscr{U}\rangle$ partially ordered by inclusion, i.e., a filter that is not properly contained in any other filter in $\langle S, \mathscr{U}\rangle$.

EXAMPLE 6-2. Let $\langle S, \mathscr{U}\rangle$ be a uniform space and $x_0 \in S$ arbitrary but fixed. Then $\mathscr{F} = \{F \subset S \mid x_0 \in F\}$ is an ultrafilter in $\langle S, \mathscr{U}\rangle$ that converges to x_0 and thus is a Cauchy filter as well.

THEOREM 6-2. *Every filter in a uniform space $\langle S, \mathscr{U}\rangle$ is contained in an ultrafilter.*

Proof. Let $\mathscr{F}$ be a filter in $\langle S, \mathscr{U}\rangle$, and let $\mathscr{C}$ be the collection of all filters that contain $\mathscr{F}$. Then $\mathscr{C}$ is nonempty and partially ordered by inclusion. Let $\mathscr{D}$ be any chain in $\mathscr{C}$, and let $\mathscr{E}$ be the collection of all sets in the filters belonging to $\mathscr{D}$. Then $\mathscr{E}$ is a filter containing all the filters in $\mathscr{D}$ and thus is an upper bound on $\mathscr{D}$. By Zorn's lemma, $\mathscr{C}$ has a maximal element $\mathscr{M}$, which is an ultrafilter containing $\mathscr{F}$.

THEOREM 6-3. *If $\mathscr{F}$ is a filter in a uniform space $\langle S, \mathscr{U}\rangle$ and $\mathscr{F}$ converges to $x_0 \in S$, then $\mathscr{F}$ accumulates at x_0.*

Proof. Let $U \in \mathscr{U}$ and $U[x_0]$ be the corresponding uniform neighborhood of x_0. Since $\mathscr{F}$ converges to x_0, $U[x_0] \in \mathscr{F}$. This implies that $U[x_0] \cap F \neq \varnothing$, $\forall F \in \mathscr{F}$, since $\mathscr{F}$ is a filter. Thus $x \in \bar{F}$, $\forall F \in \mathscr{F}$, and x is an adherent point of $\mathscr{F}$.

THEOREM 6-4. *In a uniform space, every convergent filter is a Cauchy filter, and every Cauchy filter converges to each of its adherent points.*

Proof. Let $\mathscr{F}$ be any filter in the uniform space $\langle S, \mathscr{U}\rangle$ that converges to some point $x_0 \in S$, and let $U \in \mathscr{U}$. There is a symmetric $V \in \mathscr{U}$ such that $V \circ V \subset U$. Since $\mathscr{F}$ converges to x_0, $V[x_0] \in \mathscr{F}$. If $\langle p, q\rangle \in V[x_0] \times V[x_0]$, then $\langle x_0, p\rangle \in V$ and $\langle x_0, q\rangle \in V$. This implies that $\langle p, q\rangle \in V \circ V^{-1} = V \circ V \subset U$. Hence $V[x_0] \times V[x_0] \subset U$, and $\mathscr{F}$ is a Cauchy filter.

Conversely, suppose that $\mathscr{F}$ is any Cauchy filter in the uniform space $\langle S, \mathscr{U}\rangle$ and that x_0 is any adherent point of $\mathscr{F}$. If $U[x_0]$ is any uniform neighborhood of x_0, then $\exists V \in \mathscr{U}$ such that $V \circ V \subset U$. Also, since $\mathscr{F}$ is a Cauchy filter, $\exists F \in \mathscr{F}$ such that $F \times F \subset V$. Since $x_0 \in \bar{F}$, $\forall F \in \mathscr{F}$, $\exists y \in V[x_0] \cap F$, which implies that $\langle x_0, y\rangle \in V$. Also, $y \in F$ and if $z \in F$, then $\langle y, z\rangle \in F \times F \subset V$. Hence $\langle x_0, z\rangle \in V \circ V \subset U$, which implies that $F \subset U[x_0]$ and $U[x_0] \in \mathscr{F}$. Thus $\mathscr{F}$ converges to x_0.

THEOREM 6-5. *Let $\mathscr{M}$ be an ultrafilter in the uniform space $\langle S, \mathscr{U}\rangle$. Then $\mathscr{M}$ converges to $x_0 \in S$ iff $\mathscr{M}$ accumulates at x_0.*

Proof. If $\mathscr{M}$ converges to x_0, then $\mathscr{M}$ accumulates at x_0 by Theorem 6-3. Conversely, suppose that $\mathscr{M}$ accumulates at x_0. Let $U[x_0]$ be any uniform neighborhood of x_0. By Exercise 6-3, $\exists M \in \mathscr{M}$ such that either $M \subset U[x_0]$ or $M \subset S - U[x_0]$. Since $\mathscr{M}$ accumulates at x_0, $U[x_0] \cap M \neq \varnothing$, $\forall M \in \mathscr{M}$. Hence $M \subset U[x_0]$, which implies that $U[x_0] \in \mathscr{M}$. Thus $\mathscr{M}$ converges to x_0.

Continuity of mappings of uniform spaces into uniform spaces has a characterization in terms of filterbases which is analogous to that given earlier for sequences. We leave as an exercise for the reader the proof of the statement that the image of a filterbase is a filterbase.

THEOREM 6-6. *Let $\langle S, \mathscr{U}\rangle$ and $\langle T, \mathscr{V}\rangle$ be uniform spaces. Then $f: \langle S, \mathscr{U}\rangle \to \langle T, \mathscr{V}\rangle$ is continuous at $x_0 \in S$ iff the filterbase $f(\mathscr{U}[x_0])$ converges to $f(x_0)$, where $\mathscr{U}[x_0] = \{U[x_0] \mid U \in \mathscr{U}\}$.*

Proof. f is continuous at x_0 iff $\forall V[f(x_0)]\ \exists U[x_0]$ such that $f(U[x_0]) \subset V[f(x_0)]$. This is precisely the statement that the filterbase $f(\mathscr{U}[x_0])$ converges to $f(x_0)$.

COROLLARY. *$f: \langle S, \mathscr{U}\rangle \to \langle T, \mathscr{V}\rangle$ is continuous iff the filterbase $f(\mathscr{B})$ converges to $f(x)$ for each $x \in S$ and each filterbase $\mathscr{B}$ in $\langle S, \mathscr{U}\rangle$ converging to x.*

Proof. Let f be continuous and $\mathscr{B}$ be any filterbase in $\langle S, \mathscr{U}\rangle$ converging to x. Then for each uniform neighborhood $U[x]$ $\exists B \in \mathscr{B}$ such that $B \subset U[x]$. Thus $f(B) \subset f(U[x])$. Since $f(\mathscr{U}[x])$ converges to $f(x)$, $f(\mathscr{B})$ also converges to $f(x)$. Conversely, suppose that the convergence condition of the corollary holds, and let $A \subset S$. If $x \in \bar{A}$, there is a filterbase $\mathscr{B}$ on A such that $\mathscr{B}$ converges to x. Thus the filterbase $f(\mathscr{B})$ converges to $f(x)$, which implies that $f(x) \in \overline{f(A)}$. Since $f(\bar{A}) \subset \overline{f(A)}$ $\forall A \subset S$, f is continuous.

EXERCISES

6-1. Let $\{x_n\}_{n \in I^+}$ be a sequence in a topological space $\langle S, \tau\rangle$. For each $n \in I^+$, let $B_n = \{x_k \mid k \geq n\}$. Show that the family $\{B_n \mid n \in I^+\}$ is a base for a filter in $\langle S, \tau\rangle$, which we call the *Fréchet filter* associated with $\{x_n\}_{n \in I^+}$.

6-2. Let $\mathscr{F}$ be an ultrafilter in a uniform space $\langle S, \mathscr{U}\rangle$. Show that $\bigcap\{F \mid F \in \mathscr{F}\}$ contains at most one point $x \in S$, and if $\bigcap\{F \mid F \in \mathscr{F}\} = \{x\}$, show that $\mathscr{F}$ is the family of all subsets of S that contain x (i.e., the constant ultrafilter).

6-3. Show that a filter $\mathscr{F}$ in a uniform space $\langle S, \mathscr{U}\rangle$ is an ultrafilter iff whenever $A \subset S$, either $A \in \mathscr{F}$ or $S - A \in \mathscr{F}$.

6-4. Let $\mathscr{F}$ be a filter in a uniform space $\langle S, \mathscr{U}\rangle$. Prove that $\mathscr{F}$ accumulates at $x \in S$ iff there is some filter containing $\mathscr{F}$ which converges to x.

6-5. Let $\langle S, \mathscr{U}\rangle$ be a uniform space and $A \subset S$. Show that $x \in \bar{A}$ iff there is a filterbase $\mathscr{B}$ on A which converges to x.

6-6. Show that $\langle S, \mathscr{U}\rangle$ is Hausdorff iff every convergent filter in $\langle S, \mathscr{U}\rangle$ has a unique limit.

6-7. Establish the following characterizations of compactness in a uniform space $\langle S, \mathscr{U}\rangle$:

(a) $\langle S, \mathscr{U}\rangle$ is compact iff every filter in $\langle S, \mathscr{U}\rangle$ has an adherent point.

(b) $\langle S, \mathscr{U}\rangle$ is compact iff every ultrafilter in $\langle S, \mathscr{U}\rangle$ converges.

6-8. Prove that if $f: \langle S, \mathscr{U}\rangle \to \langle T, \mathscr{V}\rangle$ is uniformly continuous and $\mathscr{B}$ is a Cauchy filterbase in $\langle S, \mathscr{U}\rangle$, then $f(\mathscr{B}) = \{f(B) \mid B \in \mathscr{B}\}$ is a Cauchy filterbase in $\langle T, \mathscr{V}\rangle$.

6-2 Completeness

We use the concepts "Cauchy sequence" and "Cauchy filter," discussed in Section 6-1, to define the notion of "completeness" for metric spaces and uniform spaces. Then we characterize compact metric spaces (uniform spaces) as those which are complete and totally bounded. Also, we show that every G_δ-subspace of a complete metric space is "topologically complete."

DEFINITION 6-6. A metric space $\langle M, d\rangle$ is *complete* iff each Cauchy sequence in $\langle M, d\rangle$ converges to some point in M.

Unfortunately, the property of completeness is not preserved by homeomorphisms, since the property of being a Cauchy sequence is not a topological property but depends upon the particular metric used. This is illustrated by the example that follows, and motivates us to define the notion of "topological completeness."

EXAMPLE 6-3. $E^1 = \langle R, d\rangle$, where $d(x, y) = |x - y|$, $\forall x, y \in R$, is complete, as we shall see shortly. Moreover, E^1 is homeomorphic to $\langle R, \rho\rangle$, where $\rho(x, y) = |x \cdot (1 + |x|)^{-1} - y \cdot (1 + |y|)^{-1}|$ $\forall x, y \in R$. Since

$$\rho(m, n) = \left|\frac{m}{1 + m} - \frac{n}{1 + n}\right| = \left|\frac{1}{1 + (1/m)} - \frac{1}{1 + (1/n)}\right|$$

$$= \frac{|1/n - 1/m|}{[1 + (1/m)][1 + (1/n)]} < \left|\frac{1}{n} - \frac{1}{m}\right| \leq \frac{1}{n} + \frac{1}{m} \leq \frac{2}{N} < \varepsilon$$

$$\forall m, n \geq N > \frac{2}{\varepsilon},$$

$\{n\}_{n \in I^+}$ is a Cauchy sequence in $\langle R, \rho\rangle$ that does not converge. Hence $\langle R, \rho\rangle$ is not complete. Also, the sequence $\{n\}_{n \in I^+}$ is not a Cauchy sequence in $\langle R, d\rangle$.

DEFINITION 6-7. $\langle M, d\rangle$ is *topologically complete* iff there is a metric ρ for M such that $\langle M, \rho\rangle$ is complete and $\langle M, \rho\rangle$ is homeomorphic to $\langle M, d\rangle$.

Completeness is not a hereditary property, since $E^1 = \langle R, d\rangle$, where $d(x, y) = |x - y|$, $\forall x, y \in R$ is complete and the subspace $R - \{0\}$ is not complete ($\{1/n\}_{n \in I^+}$ is a Cauchy sequence in $R - \{0\}$ that does not converge there). The reader is invited to show that a subspace of a complete metric space is complete iff it is closed. We shall prove a more general result than this after we have obtained Cantor's characterization of completeness.

THEOREM 6-7 (Cantor). *$\langle M, d\rangle$ is complete iff whenever $\{A_n\}_{n\in I^+}$ is a nested decreasing sequence of closed subsets of M such that $\lim_{n\to\infty} \delta(A_n) = 0$, then $\bigcap\{A_n \mid n \in I^+\}$ is a singleton.*

Proof. Assume that $\langle M, d\rangle$ is complete and that $\{A_n\}_{n\in I^+}$ is a nested decreasing sequence of closed subsets of M such that $\lim_{n\to\infty} \delta(A_n) = 0$. Let $x_n \in A_n$ $\forall n \in I^+$, and let $\varepsilon > 0$ be arbitrary. Then $\exists N \in I^+$ such that $\delta(A_n) < \varepsilon/2$, $\forall n > N$, which implies that $d(x_m, x_n) < \varepsilon$ if $m, n \geq N + 1$. Thus $\{x_n\}_{n\in I^+}$ is a Cauchy sequence and must converge to some $x \in M$. Also, the subsequence $\{x_{n+k}\}_{k\in I^+\cup\{0\}}$ must converge to x, and it is contained in A_n, $\forall n \in I^+$. Hence $x \in \overline{A}_n = A_n$, $\forall n \in I^+$, since A_n is closed, which implies that $x \in \bigcap\{A_n \mid n \in I^+\}$. If there existed a $y \in \bigcap\{A_n \mid n \in I^+\}$ such that $y \neq x$, then $\delta(\bigcap\{A_n \mid n \in I^+\}) > 0$, which contradicts our assumption that $\lim_{n\to\infty} \delta(A_n) = 0$.

Conversely, suppose that any nested decreasing sequence of closed subsets of M with diameters tending to zero has a singleton intersection. Let $\{x_n\}_{n\in I^+}$ be any Cauchy sequence in $\langle M, d\rangle$. For each $n \in I^+$, let $B_n = \{x_k \mid k \geq n\}$ and $A_n = \overline{B}_n$. Then $\{A_n\}_{n\in I^+}$ is a nested decreasing sequence of closed subsets of M with diameters tending to zero. Thus $\bigcap\{A_n \mid n \in I^+\} = \{x\}$ for some $x \in M$. If $\varepsilon > 0$ is arbitrary, then $\exists N \in I^+$ such that $\delta(B_n) < \varepsilon$ and thus $\delta(A_n) < \varepsilon$, $\forall n \geq N$. This implies that $d(x_n, x) < \varepsilon$, $\forall n \geq N$, so that $x_n \to x$. Thus $\langle M, d\rangle$ is complete.

THEOREM 6-8. *Every G_δ-subspace of a complete metric space $\langle M, d\rangle$ is topologically complete.*

Proof. Let $A = \bigcap\{G_n \mid n \in I^+\}$ be any G_δ-subspace of $\langle M, d\rangle$, where $G_n \in \tau_d$, $\forall n \in I^+$. For each $n \in I^+$, define a function $f_n : G_n \to E^1$ by

$$f_n(x) = \frac{1}{d(x, M - G_n)}.$$

Then $\forall\langle x, y\rangle \in G_n \times G_n$; let $\sigma_n(x, y)$ be given by

$$\sigma_n(x, y) = \frac{|f_n(x) - f_n(y)|}{1 + |f_n(x) - f_n(y)|}.$$

Since $\sigma_n(x, y) = 0$ does not imply that $x = y$, σ_n is not necessarily a metric. For all $x, y, z \in G_n$, we do have $\sigma_n(x, y) + \sigma_n(y, z) \geq \sigma_n(x, z)$. We define a metric ρ for A by

$$\rho(x, y) = d(x, y) + \sum_{n=1}^{\infty} 2^{-n}\sigma_n(x, y) \qquad \forall\langle x, y\rangle \in A \times A.$$

If $\{x_n\}_{n \in I^+}$ is a Cauchy sequence in $\langle A, \rho \rangle$, then $\{x_n\}_{n \in I^+}$ is also a Cauchy sequence in $\langle M, d \rangle$ and thus must d-converge to some $x \in M$. If $x \in A$, then $\{x_n\}_{n \in I^+}$ also ρ-converges to x. Thus $\langle A, d \rangle$ is topologically complete, since $\tau_d = \tau_\rho$. If $x \notin A$, then $\exists N \in I^+$ such that $x \in M - G_n$, $\forall n > N$. Fix some $k \in I^+$ and consider $\sigma_n(x_k, x_{k+j})$ for all $n > N$. Since $x_{k+j} \to x$, $\sigma_n(x_k, x_{k+j}) \to 1$ as $j \to \infty$, which implies that $d(x_{k+j}, M - G_n) \to 0$ as $j \to \infty$. Then $\lim_{j \to \infty} \rho(x_k, x_{k+j}) \geq \sum_{n=N}^{\infty} 2^{-n} = 2^{-(N-1)}$. This implies that $\{x_n\}_{n \in I^+}$ is not a Cauchy sequence in $\langle A, \rho \rangle$, which is a contradiction.

We next show that the product of a collection of complete metric spaces is a complete metric space under certain circumstances. This result has as an important corollary the completeness of E^n $\forall n \in I^+$.

THEOREM 6-9. *Let $\langle M_\alpha, d_\alpha \rangle$ be a complete metric space $\forall \alpha \in \Lambda$. Let $\langle M, \tau \rangle = \langle \prod M_\alpha, \prod \tau_{d_\alpha} \rangle$. If $\langle M, \tau \rangle$ is metrizable by a metric ρ such that $\rho(x, y) \geq d_\alpha(\pi_\alpha(x), \pi_\alpha(y))$, $\forall \alpha \in \Lambda$ and $\forall x, y \in M$, then $\langle M, \rho \rangle$ is a complete metric space.*

Proof. Let $\{A_n\}_{n \in I^+}$ be any nested decreasing sequence of nonempty closed subsets of M such that $\lim_{n \to \infty} \delta(A_n) = 0$. For each $\alpha \in \Lambda$, $\{\pi_\alpha(A_n)\}_{n \in I^+}$ is a nested decreasing sequence of nonempty closed subsets of M_α such that $\lim_{n \to \infty} \delta(\pi_\alpha(A_n)) = 0$. By Theorem 6-7, $\bigcap \{\pi_\alpha(A_n) \mid n \in I^+\}$ is a singleton $\{x_\alpha\}$ $\forall \alpha \in \Lambda$. Thus $\bigcap \{A_n \mid n \in I^+\} = \{x\}$, where $\pi_\alpha(x) = x_\alpha$ $\forall \alpha \in \Lambda$. Then, by Theorem 6-7, $\langle M, \rho \rangle$ is complete.

COROLLARY. *The following are complete: I^ω, I^n, and E^n $\forall n \in I^+$.*

Our next result characterizes compact metric spaces in terms of completeness and total boundedness.

THEOREM 6-10. *A metric space $\langle M, d \rangle$ is compact iff it is complete and totally bounded.*

Proof. Suppose that $\langle M, d \rangle$ is compact. Then $\langle M, d \rangle$ is totally bounded by Theorem 3-21 and complete, since $\langle M, d \rangle$ is also sequentially compact. Conversely, suppose that $\langle M, d \rangle$ is complete and totally bounded. Then $\forall \varepsilon > 0$ there is a finite open cover $\mathscr{U}_\varepsilon$ for M by sets of diameter less than ε. Let $\{x_n\}_{n \in I^+}$ be any sequence in M. If any element is repeated infinitely often, then $\{x_n\}_{n \in I^+}$ contains a convergent constant subsequence. Otherwise, $\forall \varepsilon > 0$ some element $U_\varepsilon \in \mathscr{U}_\varepsilon$ contains infinitely many distinct points of the set $\{x_n \mid n \in I^+\}$. Let x_{n_1} be the first element of the sequence contained in U_1 and x_{n_j} be the first element of $\{x_k \mid k > n_{j-1}\}$ contained in $U_{1/j}$ $\forall j \in I^+$. Then $\{x_{n_j}\}_{j \in I^+}$ is a Cauchy subsequence of $\{x_n\}_{n \in I^+}$ and must converge to some $x \in M$, since $\langle M, d \rangle$ is complete. Hence $\langle M, d \rangle$ is sequentially compact, which implies that it is also compact.

We conclude this section by defining the notion of completeness for uniform spaces and obtaining the analogue of Theorem 6-10 in the context of uniform spaces.

DEFINITION 6-8. A uniform space $\langle S, \mathscr{U} \rangle$ is *complete* iff every Cauchy filter in $\langle S, \mathscr{U} \rangle$ converges to a point of S.

DEFINITION 6-9. A uniform space $\langle S, \mathscr{U} \rangle$ is *totally bounded* (*precompact*) iff $\forall U \in \mathscr{U}$, there is a finite set $\{x_1, x_2, \ldots, x_n\} \subset S$ such that $S = \bigcup\{U[x_i] \mid i = 1, \ldots, n\}$.

THEOREM 6-11. *A uniform space $\langle S, \mathscr{U} \rangle$ is totally bounded iff every filter in $\langle S, \mathscr{U} \rangle$ is contained in a Cauchy filter in $\langle S, \mathscr{U} \rangle$.*

Proof. Let $\langle S, \mathscr{U} \rangle$ be totally bounded, and let $\mathscr{F}$ be any filter in $\langle S, \mathscr{U} \rangle$. Then $\mathscr{F}$ is contained in an ultrafilter $\mathscr{M}$ by Theorem 6-2. Let $U \in \mathscr{U}$. There is a symmetric $V \in \mathscr{U}$ such that $V \circ V \subset U$. Since $\langle S, \mathscr{U} \rangle$ is totally bounded, there is a finite set $\{x_1, \ldots, x_n\} \subset S$ such that $S = \bigcup\{V[x_i] \mid i = 1, \ldots, n\}$. If for each $i = 1, \ldots, n$ $\exists M_i \in \mathscr{M}$ such that $V[x_i] \cap M_i = \varnothing$, then $M = \bigcap\{M_i \mid i = 1, \ldots, n\} \in \mathscr{M}$ and

$$\begin{aligned} M \cap S &= M \cap (\bigcup\{V[x_i] \mid i = 1, \ldots, n\}) \\ &= \bigcup\{M \cap V[x_i] \mid i = 1, \ldots, n\} \\ &\subset \bigcup\{M_i \cap V[x_i] \mid i = 1, \ldots, n\} = \varnothing, \end{aligned}$$

which implies that $M = \varnothing$, a contradiction. Thus, for some i, $V[x_i] \cap M \neq \varnothing$ $\forall M \in \mathscr{M}$. Since $\mathscr{M}$ is an ultrafilter, $V[x_i] \in \mathscr{M}$. Then $\mathscr{M}$ is a Cauchy filter since $V[x_i] \times V[x_i] \subset U$.

Conversely, suppose that each filter in $\langle S, \mathscr{U} \rangle$ is contained in a Cauchy filter in $\langle S, \mathscr{U} \rangle$, and let $U \in \mathscr{U}$. Assume that $U[A] = \bigcup\{U[x] \mid x \in A\} \neq S$ for every finite subset A of S. Then $\{S - U[A] \mid A \subset S \text{ is finite}\}$ is a filter in $\langle S, \mathscr{U} \rangle$ and is contained in some Cauchy filter $\mathscr{F}$. This implies that $\exists F \in \mathscr{F}$ such that $F \times F \subset U$. Then $F \subset U[x]$ for some particular $x \in F$ so that $U[x] \in \mathscr{F}$. Since $\{x\}$ is finite, $S - U[x] \in \mathscr{F}$. However, this is a contradiction, since $U[x] \cap (S - U[x]) = \varnothing \notin \mathscr{F}$. Hence there is some finite subset A of S such that $U[A] = S$, and $\langle S, \mathscr{U} \rangle$ is totally bounded.

THEOREM 6-12. *A uniform space $\langle S, \mathscr{U} \rangle$ is compact iff it is complete and totally bounded.*

Proof. Let $\langle S, \mathscr{U} \rangle$ be compact and $U \in \mathscr{U}$. Then $\{U[x] \mid x \in S\}$ is an open cover of S and must contain a finite subcover. Thus $\langle S, \mathscr{U} \rangle$ is totally bounded. If $\mathscr{F}$ is any Cauchy filter in $\langle S, \mathscr{U} \rangle$, then $\mathscr{F}$ has an adherent point x by Exercise 6-7. Hence $\mathscr{F}$ converges to x by Theorem 6-4, and $\langle S, \mathscr{U} \rangle$ is complete. Conversely, suppose that $\langle S, \mathscr{U} \rangle$ is complete and totally bounded. If $\mathscr{F}$ is any filter in $\langle S, \mathscr{U} \rangle$, then $\mathscr{F}$ is contained in some Cauchy filter $\mathscr{G}$ by Theorem 6-11. Since $\langle S, \mathscr{U} \rangle$ is complete, $\mathscr{G}$ must converge to some $x \in S$. By Exercise 6-4, x is an adherent point of $\mathscr{F}$, which implies that $\langle S, \mathscr{U} \rangle$ is compact by Exercise 6-7.

EXERCISES

6-9. Prove that a subspace $\langle A, d \rangle$ of a complete metric space $\langle M, d \rangle$ is complete iff A is closed. Show that the irrationals are a noncomplete subspace of E^1.

6-10. Show that a metric space $\langle M, d \rangle$ is totally bounded iff every sequence in $\langle M, d \rangle$ contains a Cauchy subsequence.

6-11. Give an example of a complete metric space that is bounded but not totally bounded and hence not compact.

6-12. Show that any discrete metric space $\langle M, d \rangle$ is topologically complete. Prove that the irrationals are topologically complete.

6-13. Let $\langle M, \sigma \rangle$ and $\langle N, \rho \rangle$ be metric spaces such that $\langle N, \rho \rangle$ is complete and bounded. If N^M is the set of all continuous functions from M into N and $d(f, g) = \sup \{\rho(f(x), g(x)) \mid x \in M\}$, $\forall f, g \in N^M$, show that $\langle N^M, d \rangle$ is a complete metric space.

6-14. Show that every closed subspace of a complete uniform space is complete, and every complete subspace of a Hausdorff uniform space is closed.

★6-3 Completions of Metric and Uniform Spaces

If a metric space $\langle M, d \rangle$ is not complete, then at least one Cauchy sequence in $\langle M, d \rangle$ fails to converge relative to d. In constructing the set of real numbers from the set of rationals, one standard method uses equivalence classes of Cauchy sequences of rationals. We describe now a similar technique for "completing" an arbitrary metric or uniform space.

Let $\langle M, d \rangle$ be a metric space, and let $\mathscr{C}^*(M)$ denote the set of all bounded, continuous, real-valued functions defined on $\langle M, d \rangle$. Let $\rho(f, g) = \sup \{|f(x) - g(x)| \mid x \in M\}$, $\forall f, g \in \mathscr{C}^*(M)$. The exercise of showing that ρ is a metric for $\mathscr{C}^*(M)$ is left to the reader. We prove that $\langle \mathscr{C}^*(M), \rho \rangle$ is complete and that $\langle M, d \rangle$ can be embedded isometrically as a dense subspace of $\langle \mathscr{C}^*(M), \rho \rangle$.

THEOREM 6-13. *$\langle \mathscr{C}^*(M), \rho \rangle$ is complete.*

Proof. Let $\{f_n\}_{n \in I^+}$ be any Cauchy sequence in $\langle \mathscr{C}^*(M), \rho \rangle$, and let $\varepsilon > 0$ be arbitrary. Then $\exists N \in I^+$ such that $|f_m(x) - f_n(x)| \leq \rho(f_m, f_n) < \varepsilon$, $\forall m, n \geq N$ and $\forall x \in M$. Hence $\{f_n(x)\}_{n \in I^+}$ is a Cauchy sequence in E^1 $\forall x \in M$ and must converge to some real number $f(x)$. Now $\forall m, n \geq N$ and $\forall x \in M$, we have $-\varepsilon < f_m(x) - f_n(x) < \varepsilon$, which is equivalent to $f_n(x) - \varepsilon < f_m(x) < f_n(x) + \varepsilon$. Holding n and x fixed and letting $m \to \infty$, we get $f_n(x) - \varepsilon \leq f(x) \leq f_n(x) + \varepsilon$ or $|f_n(x) - f(x)| \leq \varepsilon$, $\forall n \geq N$ and $\forall x \in M$. This implies that $\rho(f_n, f) = \sup \{|f_n(x) - f(x)| \mid x \in M\} \leq \varepsilon$, $\forall n \geq N$. Hence $f_n \to f$, and since the convergence is uniform and f_n is continuous and bounded $\forall n \in I^+$, we have that $f \in \mathscr{C}^*(M)$. Thus $\langle \mathscr{C}^*(M), \rho \rangle$ is complete.

DEFINITION 6-10. Two metric space $\langle M, \sigma \rangle$ and $\langle N, \rho \rangle$ are *isometric* iff there exists a homeomorphism h of $\langle M, \sigma \rangle$ onto $\langle N, \rho \rangle$ such that $\sigma(x, y) = \rho(h(x), h(y))$, $\forall x, y \in M$. The homeomorphism h is called an *isometry*.

THEOREM 6-14. *Any metric space $\langle M, d \rangle$ is isometric to a dense subspace of the complete metric space $\langle \mathscr{C}^*(M), \rho \rangle$.*

Proof. Let $x_0 \in M$ be arbitrary but fixed. For each $x \in M$, we define a real-valued function $f(x)$ by setting $[f(x)](t) = d(t, x) - d(t, x_0)$, $\forall t \in M$. Since d is continuous, then $f(x)$ is also. $|[f(x)](t)| \leq d(x, x_0)$, $\forall t \in M$, as a consequence of the triangle inequality. Thus $f(x)$ is bounded $\forall x \in M$; i.e., $f(x) \in \mathscr{C}^*(M)$, $\forall x \in M$. Moreover, $\rho(f(x), f(y)) = \sup \{|[f(x)](t) - [f(y)](t)| \mid t \in M\} = \sup \{|d(t, x) - d(t, y)| \mid t \in M\} \geq |d(t, x) - d(t, y)|$, $\forall t \in M$. For $t = y$ this yields $\rho(f(x), f(y)) \geq d(y, x) = d(x, y)$. If $d(t, x) - d(t, y) > d(x, y)$ for some $t \in M$, then $d(t, x) > d(t, y) + d(x, y) \geq d(t, x)$ gives us a contradiction. Also, if $d(t, x) - d(t, y) < -d(x, y)$ for some $t \in M$, then $d(y, t) \leq d(x, y) + d(t, x) < d(t, y)$ gives us a contradiction. Hence $|d(t, x) - d(t, y)| \leq d(x, y)$, $\forall t \in M$, which implies that $\rho(f(x), f(x)) \leq d(x, y)$. Thus $\rho(f(x), f(y)) = d(x, y)$, which shows that $\langle M, d \rangle$ is isometric to the subspace $f(M)$, which is dense in the complete subspace $\overline{f(M)}$ of $\langle \mathscr{C}^*(M), \rho \rangle$.

The constructive proof of Theorem 6-14 would proceed as follows. Two Cauchy sequences $\{x_n\}_{n \in I^+}$ and $\{y_n\}_{n \in I^+}$ in $\langle M, d \rangle$ are "equivalent" iff $\lim_{n \to \infty} d(x_n, y_n) = 0$. This "equivalence" relation partitions the collection of all Cauchy sequences in $\langle M, d \rangle$ into disjoint equivalence classes which we take to be the points of a new space M^* having a metric d^* given by $d^*(x^*, y^*) = \lim_{n \to \infty} d(x_n, y_n)$, where $\{x_n\}_{n \in I^+}$ and $\{y_n\}_{n \in I^+}$ are representatives, respectively, of the equivalence classes x^* and y^* of Cauchy sequences in $\langle M, d \rangle$. One defines a mapping f of $\langle M, d \rangle$ into $\langle M^*, d^* \rangle$ by letting $f(x)$ be the equivalence class of all Cauchy sequences in $\langle M, d \rangle$ having x as the limit for each $x \in M$. Then one shows that f is an isometry of $\langle M, d \rangle$ onto $\langle f(M), d^* \rangle$ and that $\langle f(M), d^* \rangle$ is a dense subspace of $\langle M^*, d^* \rangle$. Details are left to the reader as an exercise but can be found in Pervin, *Foundations of General Topology*, pp. 123–24.

Because of the duality that exists between the notion of a Cauchy sequence in a metric space and that of a Cauchy filter in a uniform space, the same sort of construction as just outlined above can be used to "complete" an arbitrary uniform space. Details of the construction can be found in Bourbaki, *Elements of Mathematics* (Part 1), pp. 191–93. As a result, we obtain the following analogue of Theorem 6-14.

THEOREM 6-15. *Every uniform space is uniformly isomorphic to a dense subspace of a complete uniform space.*

EXERCISES

6-15. Show that $\langle \mathscr{C}^*(M), \rho \rangle$ is a metric space.

6-16. Show that the relation "$\sim$" given by $\{x_n\}_{n \in I^+} \sim \{y_n\}_{n \in I^+}$ iff $\lim_{n \to \infty} d(x_n, y_n) = 0$ is an equivalence relation on the collection of all Cauchy sequences in a metric space $\langle M, d \rangle$.

6-17. Supply the details to show that $\langle M, d \rangle$ is isometric to $\langle f(M), d^* \rangle$ and that $\langle f(M), d^* \rangle$ is dense in $\langle M^*, d^* \rangle$.

★6-4 Applications of Completeness

We conclude our discussion of completeness by proving two important theorems that are valid for complete metric spaces. One is the Baire category theorem and the other is the Banach contraction mapping theorem. We preface our proof of the Baire category theorem with a brief discussion of "Baire spaces."

DEFINITION 6-11. A topological space $\langle S, \tau \rangle$ is a *Baire space* iff the intersection of each countable family of open dense subsets of S is dense in S.

THEOREM 6-16. *Let $\langle S, \tau \rangle$ be a Baire space. If $\mathscr{C} = \{F_n \mid n \in I^+\}$ is any countable closed cover of S, then some member $F_n \in \mathscr{C}$ must contain a member of $\tau - \{\varnothing\}$; i.e., the interior of F_n is nonempty for some $n \in I^+$.*

Proof. Since $S = \bigcup\{F_n \mid n \in I^+\}$, we have $\varnothing = \bigcap\{S - F_n \mid n \in I^+\}$. Moreover, since $\langle S, \tau \rangle$ is a Baire space, some set $S - F_n$ cannot be dense in S; i.e., $\overline{S - F_n} \neq S$, which implies that the interior of F_n $(= S - \overline{S - F_n})$ is nonempty for some $n \in I^+$.

If $\langle S, \tau \rangle$ is any topological space and $A \subset S$, recall from Definition 1-11 that A is *nowhere dense* in S iff the interior of $\bar{A}$ is empty; i.e., $S - \overline{S - \bar{A}} = \varnothing$ or $S - \bar{A}$ is dense in S.

DEFINITION 6-12. $A \subset S$ is of the *first category* (*meager*) iff A is the union of a countable number of nowhere dense subsets of S. Otherwise, A is of the *second category*.

EXAMPLE 6-4. In E^1 the set of rationals is of the first category, since it is countable and the singletons are nowhere dense in E^1. Also, the complementary set of irrationals is of the second category.

THEOREM 6-17. *Let $\langle S, \tau \rangle$ be a Baire space, and let $A \subset S$ be of the first category. Then A has an empty interior.*

Proof. Let $A = \bigcup\{A_n \mid n \in I^+\}$, where A_n is nowhere dense in S $\forall n \in I^+$, and let $G \in \tau$ such that $G \subset A = \bigcup\{A_n \mid n \in I^+\} \subset \bigcup\{\overline{A_n} \mid n \in I^+\}$. Then $\bigcap\{S - \overline{A_n} \mid n \in I^+\} \subset S - G$. Since $S - \overline{A_n} \in \tau$ and dense in S, $\forall n \in I^+$, $S - G$ is also dense in S. Thus $\overline{S - G} = S - G = S$, which implies that $G = \varnothing$, so that A has an empty interior.

THEOREM 6-18 (Baire Category Theorem). *Every complete metric space $\langle M, d \rangle$ is a Baire space and thus has the properties described in Theorems 6-16 and 6-17.*

Proof. Let $\{G_n \mid n \in I^+\}$ be any countable family of open dense subsets of M, and let $U \in \tau - \{\varnothing\}$. Since G_1 is dense, $U \cap G_1 \neq \varnothing$. Hence there is some open d-sphere B_1 with $\delta(B_1) \leq 1$ such that $\overline{B_1} \subset U \cap G_1$. Proceeding inductively, we obtain a sequence $\{B_n\}_{n \in I^+}$ of open d-spheres such that $\delta(B_n) \leq 1/n$ and $\overline{B_n} \subset B_{n-1} \cap G_n$ for each $n > 1$. Since $\langle M, d \rangle$ is complete and $\{\overline{B_n} \mid n \in I^+\}$ is a nested decreasing sequence of closed sets with $\lim_{n \to \infty} \delta(B_n) = 0$, we have $\bigcap\{\overline{B_n} \mid n \in I^+\} = \{x\}$ for some $x \in M$ by

Theorem 6-7. This implies that $\bigcap\{G_n \mid n \in I^+\} \neq \varnothing$, and $\langle M, d\rangle$ is a Baire space.

As a second application of completeness, we define the notion of a contraction mapping and show that every contraction mapping on a complete metric space has a unique fixed point. This result is the famous Contraction Mapping Theorem of Banach and has been generalized in many ways in recent years.

DEFINITION 6-13. Let $\langle M, d\rangle$ be a metric space and let $f: \langle M, d\rangle \to \langle M, d\rangle$ be any function. Then $x_0 \in M$ is a *fixed point* under f iff $f(x_0) = x_0$.

EXAMPLE 6-5. The function $f: E^1 \to E^1$ given by $f(x) = x$, $\forall x \in R$, leaves every point of R fixed. However, the function $g: E^1 \to E^1$ given by $g(x) = x + 1$ leaves no point of R fixed. In between these two extremes, a function such as $h(x) = -x$ leaves exactly one real number fixed (zero).

DEFINITION 6-14. Let $\langle M, d\rangle$ be a metric space. Then $f: \langle M, d\rangle \to \langle M, d\rangle$ is a *contraction* (relative to d) iff $\exists \alpha < 1$ such that $d(f(x), f(y)) \leq \alpha d(x, y)$, $\forall x, y \in M$.

THEOREM 6-19 (Banach). *Let $\langle M, d\rangle$ be a complete metric space, and let $f: \langle M, d\rangle \to \langle M, d\rangle$ be a contraction (relative to d). Then f has a unique fixed point.*

Proof

Uniqueness. If $f(x_0) = x_0$ and $f(y_0) = y_0$, then $d(x_0, y_0) = d(f(x_0), f(y_0)) \leq \alpha d(x_0, y_0) < d(x_0, y_0)$. This implies that $d(x_0, y_0) = 0$, so that $x_0 = y_0$.

Existence. Let $x \in M$ be arbitrary. Consider the sequence $\{x, f(x), f(f(x)) = f^2(x), \ldots, f^n(x), \ldots\}$ of iterates of f. We have

$$d(f(x), f^2(x)) \leq \alpha d(x, f(x))$$
$$d(f^2(x), f^3(x)) \leq \alpha d(f(x), f^2(x)) \leq \alpha^2 d(x, f(x))$$

and, by induction, that $d(f^n(x), f^{n+1}(x)) \leq \alpha^n d(x, f(x))$. Then, for any particular $n \in I^+$ and $\forall m > n$, we get

$$\begin{aligned} d(f^n(x), f^m(x)) &\leq d(f^n(x), f^{n+1}(x)) + \cdots + d(f^{m-1}(x), f^m(x)) \\ &\leq (\alpha^n + \alpha^{n+1} + \cdots + \alpha^{m-1}) d(x, f(x)) \\ &= (1 + \alpha + \cdots + \alpha^{m-n-1}) \alpha^n d(x, f(x)) \\ &< \frac{\alpha^n}{1 - \alpha} \cdot d(x, f(x)), \end{aligned}$$

since $\sum_{k=0}^{\infty} \alpha^k$ is a geometric series with ratio $\alpha < 1$. Since $\alpha^n \to 0$ as $n \to \infty$, we see that $\{f^n(x)\}_{n \in I^+ \cup \{0\}}$ is a Cauchy sequence in $\langle M, d \rangle$ and must therefore converge to some $x_0 \in M$. Moreover, $f^n(x) \to x_0$ as $n \to \infty$ implies that $f(f^n(x)) = f^{n+1}(x) \to x_0$ as $n \to \infty$, since every subsequence of a convergent sequence must converge to the same limit. Since f is also continuous (why?), $f^{n+1}(x) \to f(x_0)$ as $n \to \infty$. Finally, since sequentially limits are unique in any Hausdorff space, we must have $f(x_0) = x_0$.

EXERCISES

6-18. Prove that any open subspace of a Baire space is a Baire space.

6-19. Show that the property of being a Baire space is invariant under continuous open surjections.

6-20. Show that $\langle R, \mathscr{L} \rangle$ is a Baire space.

6-21. Prove that any locally compact Hausdorff space is a Baire space.

6-22. Show that any contraction $f: \langle M, d \rangle \to \langle M, d \rangle$ is necessarily continuous.

6-23. Prove that if $f: \langle M, d \rangle \to \langle M, d \rangle$ is a contraction on a complete a-metric space $\langle M, d \rangle$, then f has a unique fixed point. (The case $a = 0$ is exactly the Contraction Mapping Theorem of Banach proved above.)

CHAPTER 7
Homotopy Theory

7-1 Retracts

In Chapters 7 and 8 we turn our attention away from "point-set" topology and look at topological spaces in terms of certain algebraic groups associated with them. These groups are topological invariants in the sense that isomorphic groups are associated with homeomorphic spaces. In this chapter we introduce the notion of homotopic mappings and define and investigate the homotopy groups associated with a topological space. In Chapter 8 we define and investigate a second sequence of groups known as the singular homology groups and obtain simple proofs of the Fundamental Theorem of Algebra and the Brouwer fixed-point theorem using the tools of homotopy and homology. We preface our discussion of homotopy theory with some remarks about "retracts." Here "mapping" means a continuous function.

DEFINITION 7-1

(1) $A \subset S$ is a *retract* of S iff there exists a mapping $r : S \to A$ such that $r(x) = x$ for each $x \in A$; i.e., the identity mapping on A can be extended continuously to S. The mapping r is called a *retraction.*

(2) $A \subset S$ is a *neighborhood retract* of S iff there exists $U \supset A$ such that $U \subset S$ is open and A is a retract of U; i.e., the identity mapping on A can be extended continuously to U.

DEFINITION 7-2

(1) A is an *absolute retract* iff given a normal space S having a closed subspace A' homeomorphic to A, then A' is a retract of S.

(2) A is an *absolute neighborhood retract* iff given a normal space S having a closed subspace A' homeomorphic to A, then A' is a neighborhood retract of S.

As consequences of the Tietze extension theorem, we can show that I^n is an absolute retract and S^n is an absolute neighborhood retract.

THEOREM 7-1. *I^n is an absolute retract.*

Proof. Let $h(I^n) = A'$, where h is a homeomorphism of I^n into a normal space S. Since A' is closed, Tietze's extension theorem implies that the mapping $h^{-1} : A' \to I^n$ has a continuous extension $g : S \to I^n$. Thus $hg \mid A'$ is the identity on A' and A' is a retract of S.

THEOREM 7-2. *If $f: A \to S^n$ is continuous, where A is a closed subspace of a normal space S, then there exists $U \supset A$ and an extension $f^*: U \to S^n$ of f to U, where $U \subset S$ is open.*

Proof. Since S^n is homeomorphic to the boundary of I^{n+1}, Tietze's extension theorem implies that the mapping $f: A \to S^n$ has a continuous extension $g: S \to I^{n+1}$. Let $p \in I^{n+1}$ with each coordinate equal to $\frac{1}{2}$. The radial projection from p is a retraction r of $I^{n+1} - \{p\}$ onto the boundary of I^{n+1}. Moreover, $g^{-1}(I^{n+1} - \{p\}) = S - g^{-1}(\{p\}) = U \supset A$ and U is open in S. Finally, if we set $f^* = rg$, then $f^*(x) = rg(x) = rf(x) = f(x)$ for each $x \in A$, which implies f^* is an extension of f to U.

EXERCISES

7-1. Show that S^n is an absolute neighborhood retract.

7-2. If A is a retract of S and $f: A \to T$ is a mapping, show that f can be extended to all of S.

7-2 Homotopic Mappings

In this section we introduce the notion of "homotopic mappings" and show that "homotopy" is an equivalence relation on the function space T^S of all mappings of a topological space S into a topological space T. The resulting equivalence classes are called "homotopy classes" and constitute a decomposition of T^S. The homotopy classes of T^S may be characterized as the arcwise-connected components of T^S.

DEFINITION 7-3. Let S and T be topological spaces. If f and g are mappings of S into T, then f and g are *homotopic* ($f \simeq g$) iff there exists a mapping $h: S \times I^1 \to T$ such that $h(x, 0) = f(x)$ and $h(x, 1) = g(x)$ for all $x \in S$. The mapping h is called a *homotopy* between f and g.

THEOREM 7-3. *Homotopy is an equivalence relation on the function space T^S of all mappings of S into T.*

Proof

(1) $\simeq$ is reflexive. If $f \in T^S$, let $h(x, t) = f(x)$ for each $t \in I^1$ and for all $x \in S$. Thus $h(x, 0) = f(x) = h(x, 1)$ and $f \simeq f$.

(2) $\simeq$ is symmetric. If $f \simeq g$, then there exists a homotopy $h: S \times I^1 \to T$ with $h(x, 0) = f(x)$ and $h(x, 1) = g(x)$ for all $x \in S$. Let $h^*(x, t) = h(x, 1 - t)$. Since $h^*(x, 0) = h(x, 1) = g(x)$ and $h^*(x, 1) = h(x, 0) = f(x)$ for all $x \in S$, we have $g \simeq f$.

(3) $\simeq$ is transitive. If $f \simeq g$ and $g \simeq k$, then there exist homotopies $h_1, h_2 : S \times I^1 \to T$ with $h_1(x, 0) = f(x)$, $h_1(x, 1) = g(x) = h_2(x, 0)$, and $h_2(x, 1) = k(x)$ for all $x \in S$. Let $h(x, t) = h_1(x, 2t)$ if $0 \le t \le \frac{1}{2}$ and $h(x, t) = h_2(x, 2t - 1)$ if $\frac{1}{2} \le t \le 1$. One easily verifies that h is continuous and that $h(x, 0) = h_1(x, 0) = f(x)$ and $h(x, 1) = h_2(x, 1) = k(x)$ for all $x \in S$. Thus $f \simeq k$.

DEFINITION 7-4. The equivalence classes $[f]$ determined by the homotopy relation on the function space T^S of all mappings $f : S \to T$ are called the *homotopy classes* of T^S.

LEMMA 7-1. *Let S be a normal space, A a closed subset of S, and $U \supset A$ an open subset of S. If $f : (U \times I^1) \cup (S \times \{0\}) \to T$ is continuous, then f has an extension f^* to $S \times I^1$ such that $f^*(x, t) = f(x, t)$ for all $\langle x, t \rangle \in (A \times I^1) \cup (S \times \{0\})$.*

Proof. Urysohn's lemma implies the existence of a mapping $\varphi : S \to I^1$ with $\varphi(A) = 1$ and $\varphi(S - U) = 0$. Let $f^*(x, t) = f(x, t \cdot \varphi(x))$. f^* is continuous, since f and φ are. If $x \in A$, then $f^*(x, t) = f(x, t)$, and if $x \in S$, then $f^*(x, 0) = f(x, 0)$.

THEOREM 7-4 (Borsuk). *Let $f \simeq g : A \to S^n$, where A is a closed subset of a separable metric space M. If f has an extension f^* to M, then g has an extension g^* to M such that $f^* \simeq g^*$.*

Proof. Since $f \simeq g$, there exists a homotopy $h : A \times I^1 \to S^n$ with $h(x, 0) = f(x)$ and $h(x, 1) = g(x)$ for all $x \in A$. Let $C = (A \times I^1) \cup (M \times \{0\}) \subset M \times I^1$, and note that C is closed. Let $F(x, 0) = f^*(x)$ if $x \in M$ and $F(x, t) = h(x, t)$ if $\langle x, t \rangle \in A \times I^1$. F is continuous, since f and h are and $h(x, 0) = f(x) = f^*(x)$ for all $x \in A$. Theorem 7-2 guarantees the existence of an open subset U of $M \times I^1$ such that $U \supset C$ and an extension F^* of F to U. There exists an open subset V of M such that $A \subset V$ and $V \times I^1 \subset U$. Thus Lemma 7-1 implies the existence of a mapping $h^* : M \times I^1 \to S^n$ such that $h^*(x, t) = F^*(x, t)$ for all $\langle x, t \rangle \in C$. Let $g^*(x) = h^*(x, 1)$. Thus $g^*(x) = F^*(x, 1) = h(x, 1) = g(x)$ for all $x \in A$, which implies that g^* is an extension of g to M and h^* is a homotopy between f^* and g^*.

COROLLARY 7-1. *If f is a mapping of a closed subset A of a separable metric space M into S^n and f is homotopic to a constant mapping, then f has an extension f^* to M that is homotopic to a constant mapping.*

Proof. Immediate from Theorem 7-4, since any constant mapping has the trivial constant extension.

EXERCISES

7-3. Show that any two paths $f, g : I^1 \to S$ on an arcwise-connected topological space S are homotopic.

7-4. Show that the homotopy classes of T^S (with the compact-open topology) are its arcwise-connected components.

7-3 Contractible and Starlike Spaces

Whether or not two mappings $f, g : S \to T$ are homotopic depends upon the spaces S and T as well as the mappings. Some spaces T have the property that all mappings of an arbitrary space S into T are homotopic. One such class of spaces, the collection of "contractible" spaces, is studied in the discussion that follows, along with its important subcollection of "starlike" metric spaces.

DEFINITION 7-5. T is *contractible* iff there exists a point $p \in T$ such that the identity mapping i on T is homotopic to the constant mapping $c : T \to \{p\}$.

DEFINITION 7-6. A metric space $\langle M, d \rangle$ is *starlike* iff there exists $p \in M$ with the property that each point $x \in M - \{p\}$ can be joined to p by a unique arc congruent to an interval. The property of being starlike is not topological, since it depends upon the metric d.

THEOREM 7-5. *If $\langle M, d \rangle$ is starlike, then $\langle M, d \rangle$ is contractible.*

Proof. Since $\langle M, d \rangle$ is starlike, there exists $p \in M$ with the property that each $x \in M - \{p\}$ can be joined to p by a unique arc px congruent to an interval. Let $h(x, t) = y \in px$ such that $d(p, y) = t \cdot d(p, x)$ for each $t \in I^1$ and all $x \in M$. Now $h(x, 0) = p$ and $h(x, 1) = x$, $\forall x \in M$, and h intuitively shrinks radially all arcs px to p. Thus $\langle M, d \rangle$ is contractible.

THEOREM 7-6. *If T is contractible, then each mapping $f: S \to T$ is homotopic to a constant mapping $cf: S \to \{p\}$ and T^S has only the single homotopy class $[cf]$.*

Proof. Let $f: S \to T$ and let i be the identity mapping on T. Since T is contractible, $i \simeq c: T \to \{p\}$ and $cf: S \to \{p\}$. Thus there exists a homotopy $h: T \times I^1 \to T$ with $h(y, 0) = y$ and $h(y, 1) = p$ for all $y \in T$. Let $\tilde{h}(x, t) = h(f(x), t)$ for $\langle x, t \rangle \in S \times I^1$. Since f and h are continuous,

$\tilde{h}$ is continuous. Moreover, $\tilde{h}(x, 0) = h(f(x), 0) = f(x)$ and $\tilde{h}(x, 1) = h(f(x), 1) = p$ and $f \simeq cf$.

EXAMPLE 7-1. I^n and I^ω are starlike and hence contractible. Thus any mapping $f: S \to I^n$ (or I^ω) is homotopic to a constant mapping by Theorem 7-6. Moreover, any mapping f of a compact space S into E^n (or H) is homotopic to a constant mapping.

EXAMPLE 7-2. $E^2 - \{0\}$ is not contractible. To see this, let $c : S^1 \to \{p\}$, where $p \in E^2 - \{0\}$ is arbitrary, and let $f: S^1 \to J$, where J is any Jordan curve enclosing the origin 0, as in Figure 7-1. Clearly, c and f are not homotopic, since J cannot be shrunk to $\{p\}$ in $E^2 - \{0\}$. Thus the contrapositive of Theorem 7-6 yields the noncontractibility of $E^2 - \{0\}$.

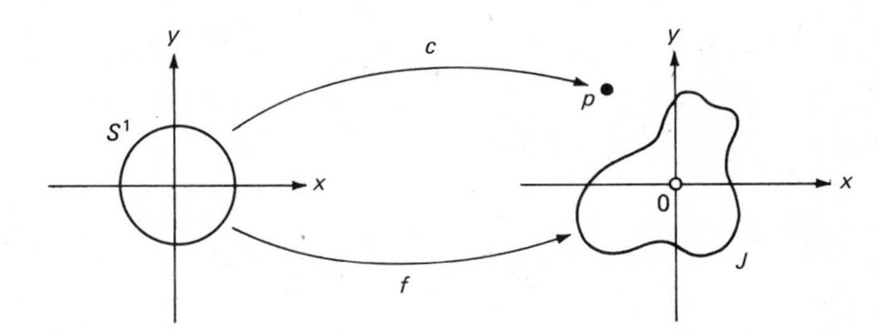

FIGURE 7-1

THEOREM 7-7. *If A is a retract of a contractible space S, then A is contractible.*

Proof. Let $r : S \to A$ be the retraction. Since S is contractible the identity mapping i on S is homotopic to the constant mapping $c : S \to \{p\}$ for some $p \in S$; i.e., there exists $h : S \times I^1 \to S$ with $h(x, 0) = x$ and $h(x, 1) = p$ for all $x \in S$. Let $\tilde{h}(x, t) = r[h(x, t)] \in A$ for $\langle x, t \rangle \in A \times I^1$. Since h and r are continuous, $\tilde{h}$ is continuous. Moreover, $\tilde{h}(x, 0) = r[h(x, 0)] = r(x) = x$ and $\tilde{h}(x, 1) = r[h(x, 1)] = r(p)$ for each $x \in A$. Thus the identity mapping $r \mid A \simeq \tilde{c} : A \to r(p)$, and A is contractible.

EXERCISES

7-5. Let $i : S^2 \to E^3$ be the inclusion mapping and $j(x) = 0$ (origin in E^3) for every $x \in S^2$. Show that $i \simeq j$.

7-6. Let S be the annulus in E^2 determined by the inequalities $a \leq x_1^2 + x_2^2 \leq b$. Let i be the identity mapping on S, and for each $x = \langle x_1, x_2 \rangle \in S$, let $j(x) = x'$, where x' is the point of intersection of the segment $0x$ and the inner boundary of S. Show that $i \simeq j$.

7-7. Show that any compact metric space that is an absolute retract is contractible.

7-8. If $f: S \to S^n$ is a mapping and $S^n - f(S) \neq \varnothing$, show that f is homotopic to a constant mapping $c: S \to \{p\}$, where $p \in S^n$.

7-4 Homotopically Equivalent Spaces

Next, we introduce the notion of two spaces being "homotopically equivalent." Any two homeomorphic spaces are homotopically equivalent, but homotopically equivalent spaces need not be homeomorphic. Also, we introduce the concepts "mapping cylinder" and "deformation retract" and show that the mapping cylinder T_f of $f: S \to T$ is homotopically equivalent to T and that T is a deformation retract of T_f.

DEFINITION 7-7. Two topological spaces S and T are *homotopically equivalent* iff there exist mappings $f: S \to T$ and $g: T \to S$ such that $fg \simeq i_T$ and $gf \simeq i_S$.

DEFINITION 7-8. Let $f: S \to T$ be a mapping and in $(S \times I^1) \cup T$, let $\pi(y) = y$ for each $y \in T$, $\pi(x, t) = \langle x, t \rangle$ if $x \in S$, $0 \leq t < 1$, and $\pi(x, 1) = f(x)$ for each $x \in S$. Let $T_f = \pi[(S \times I^1) \cup T]$ and let $G \subset T_f$ be open iff $\pi^{-1}(G)$ is open in $S \times I^1$. The resulting topological space T_f is called the *mapping cylinder* of f.

DEFINITION 7-9. $D \subset S$ is a *deformation retract* of S iff there exists a retraction r of S onto D and a homotopy $h: S \times I^1 \to S$ with $h(x, 0) = x$ and $h(x, 1) = r(x)$ for all $x \in S$ and such that $h(x, t) = x$ for all $x \in D$ and $t \in I^1$.

THEOREM 7-8. *If $f: S \to T$ is continuous, then T_f is homotopically equivalent to T.*

Proof. Let $g(x, t) = f(x)$ if $\langle x, t \rangle \in S \times I^1$ and $g(y) = y$ if $y \in T$. Since f is continuous, g is a continuous function of T_f onto T. Let $i: T \to T_f$ be the identity injection mapping, so that $gi(y) = g(y) = y$ and $gi = i_T$. Moreover, $ig(y) = i(y) = y$ for each $y \in T$ and $ig(x, t) = i(f(x)) = f(x)$ for

$\langle x, t \rangle \in S \times I^1$. Let $h(y, s) = y$ if $y \in T$ and $s \in I^1$, and let $h(\langle x, t \rangle, s) = \langle x, (1 - s)t + s \rangle$ if $\langle x, t \rangle \in S \times I^1$ and $s \in I^1$. Thus h is a continuous mapping of $T_f \times I^1$ onto T_f. Also, $h(y, 0) = y$ and $h(\langle x, t \rangle, 0) = \langle x, t \rangle$, which implies that $h(z, 0) = z$ for each $z \in T_f$, and $h(y, 1) = y$ and $h(\langle x, t \rangle, 1) = \langle x, 1 \rangle = f(x)$, which implies that $h(z, 1) = ig(z)$ for each $z \in T_f$. Hence $ig \simeq i_{T_f}$.

COROLLARY 7-2. *If $f : S \to T$ is continuous, then T is a deformation retract of T_f.*

Proof. Let $g : T_f \to T$ and $h : T_f \times I^1 \to T_f$ be as in the proof of Theorem 7-8. Since $g(y) = y$ for all $y \in T$, g is a retraction of T_f onto T. The mapping h is a homotopy between i_{T_f} and g. Moreover, $h(y, s) = y$ for each $y \in T$ so that T is a deformation retract of T_f.

EXERCISE

7-9. Show that S^1 and $S^1 \times I^1$ are homotopically equivalent.

7-5 The Fundamental Group

In this section we introduce the relation "homotopy modulo x_0" in the collection $C(S, x_0)$ of all closed paths on S based at $x_0 \in S$. This is shown to be an equivalence relation on $C(S, x_0)$, and the resulting equivalence (homotopy) classes form a group $\prod_1 (S, x_0)$ called the fundamental group of S modulo x_0. If S is arcwise connected, $\prod_1 (S, x_0)$ is shown to be independent of the base point x_0. Also, it is shown that if $\langle S, \{x_0\} \rangle$ and $\langle T, \{y_0\} \rangle$ are homotopically equivalent, then $\prod_1 (S, x_0) \simeq \prod_1 (T, y_0)$. From this result it follows that any two homeomorphic spaces have isomorphic fundamental groups.

DEFINITION 7-10. If $f : I^1 \to S$ is continuous, then f is a *closed path on S based at $x_0 \in S$* iff $f(0) = f(1) = x_0$. For each $x_0 \in S$, let $C(S, x_0)$ denote the collection of all closed paths on S based at x_0 with the compact-open topology.

DEFINITION 7-11. If $f, g \in C(S, x_0)$, then f is *homotopic* to g *modulo* x_0 ($f \tilde{\simeq}_{x_0} g$) iff there exists a mapping $h : I^2 \to S$ such that $h(s, 0) = f(s)$ and $h(s, 1) = g(s)$ for all $s \in I^1$, and $h(0, t) = h(1, t) = x_0$ for all $t \in I^1$. See Figure 7-2.

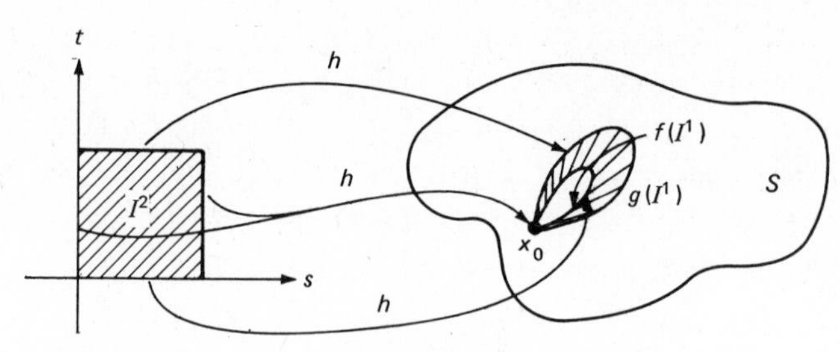

FIGURE 7-2

THEOREM 7-9. *Homotopy modulo* x_0 *is an equivalence relation on* $C(S, x_0)$.

Proof

(1) $\tilde{\tilde{x}}_0$ is reflexive. If $f \in C(S, x_0)$, let $h(s, t) = f(s)$ for each $t \in I^1$ and all $s \in I^1$. Thus $h(s, 0) = f(s) = h(s, 1)$ for all $s \in I^1$, and $h(0, t) = f(0) = x_0 = f(1) = h(1, t)$ for all $t \in I^1$. Hence $f \tilde{\tilde{x}}_0 f$.

(2) $\tilde{\tilde{x}}_0$ is symmetric. $f \tilde{\tilde{x}}_0 g$ implies the existence of a homotopy $h : I^2 \to S$ with $h(s, 0) = f(s)$ and $h(s, 1) = g(s)$ for all $s \in I^1$, and $h(0, t) = h(1, t) = x_0$ for all $t \in I^1$. Let $h^*(s, t) = h(s, 1 - t)$. Since $h^*(s, 0) = h(s, 1) = g(s)$ and $h^*(s, 1) = h(s, 0) = f(s)$ for all $s \in I^1$, and $h^*(0, t) = h(0, 1 - t) = x_0 = h(1, 1 - t) = h^*(1, t)$ for all $t \in I^1$, we have $g \tilde{\tilde{x}}_0 f$.

(3) $\tilde{\tilde{x}}_0$ is transitive. $f \tilde{\tilde{x}}_0 g$ and $g \tilde{\tilde{x}}_0 k$ imply the existence of homotopies $h_1, h_2 : I^2 \to S$ with $h_1(s, 0) = f(s)$, $h_1(s, 1) = g(s) = h_2(s, 0)$, and $h_2(s, 1) = k(s)$ for all $s \in I^1$, and $h_1(0, t) = h_1(1, t) = x_0 = h_2(0, t) = h_2(1, t)$ for all $t \in I^1$. Let $h(s, t) = h_1(s, 2t)$ if $0 \le t \le \frac{1}{2}$, and $h(s, t) = h_2(s, 2t - 1)$ if $\frac{1}{2} \le t \le 1$. h is continuous, since h_1 and h_2 are. Moreover, $h(s, 0) = h_1(s, 0) = f(s)$ and $h(s, 1) = h_2(s, 1) = k(s)$ for all $s \in I^1$, and $h(0, t) = h(1, t) = x_0$ for all $t \in I^1$. Thus $f \tilde{\tilde{x}}_0 k$.

DEFINITION 7-12. The equivalence classes $[f]$ determined by homotopy modulo x_0 on the collection $C(S, x_0)$ of all closed paths f on S based at $x_0 \in S$ are called the homotopy classes of $C(S, x_0)$. The collection of these homotopy classes is denoted by $\prod_1 (S, x_0)$.

DEFINITION 7-13

(1) If $f, g \in C(S, x_0)$, we define the *juxtaposition* $f * g$ of f and g as follows:

$$(f * g)(s) = \begin{cases} f(2s) & \text{if } 0 \le s \le \frac{1}{2} \\ g(2s - 1) & \text{if } \frac{1}{2} \le s \le 1. \end{cases}$$

Thus $f * g \in C(S, x_0)$ and $*$ is a binary operation on $C(S, x_0)$.

(2) If $[f], [g] \in \prod_1 (S, x_0)$, let $[f] \circ [g] = [f * g]$. (See Exercise 7-10.)

THEOREM 7-10. *$\prod_1 (S, x_0)$ is a group with respect to "$\circ$."*

Proof

(1) "$\circ$" is associative. In view of Definition 7-13(2), we need only show that $(f * g) * k \;\tilde{\tilde{x}}_0\; f * (g * k)$ for $f, g, k \in C(S, x_0)$. By Definition 7-13(1), we have

$$[(f * g) * k](s) = \begin{cases} f(4s) & \text{if } 0 \leq s \leq \frac{1}{4} \\ g(4s - 1) & \text{if } \frac{1}{4} \leq s \leq \frac{1}{2} \\ k(2s - 1) & \text{if } \frac{1}{2} \leq s \leq 1 \end{cases}$$

and

$$[f * (g * k)](s) = \begin{cases} f(2s) & \text{if } 0 \leq s \leq \frac{1}{2} \\ g(4s - 2) & \text{if } \frac{1}{2} \leq s \leq \frac{3}{4} \\ k(4s - 3) & \text{if } \frac{3}{4} \leq s \leq 1. \end{cases}$$

We define a homotopy between $(f * g) * k$ and $f * (g * k)$ as follows:

$$h(s, t) = \begin{cases} f\left(\dfrac{4s}{1 + t}\right) & \text{if } \langle s, t\rangle \in I^2 \text{ and } t \geq 4s - 1 \\ g(4s - t - 1) & \text{if } \langle s, t\rangle \in I^2 \text{ and } 4s - 1 \geq t \geq 4s - 2 \\ k\left(\dfrac{4s - t - 2}{2 - t}\right) & \text{if } \langle s, t\rangle \in I^2 \text{ and } 4s - 2 \geq t. \end{cases}$$

Figure 7-3 illustrates this situation. One easily verifies that $h : I^2 \to S$ is continuous, since f, g, and k are continuous. Moreover, the following is true:

$$h(s, 0) = \begin{cases} f(4s) & \text{if } 0 \geq 4s - 1 \text{ (i.e., } 0 \leq s \leq \frac{1}{4}) \\ g(4s - 1) & \text{if } 4s - 1 \geq 0 \geq 4s - 2 \text{ (i.e., } \frac{1}{4} \leq s \leq \frac{1}{2}) \\ k(2s - 1) & \text{if } 4s - 2 \geq 0 \text{ (i.e., } \frac{1}{2} \leq s \leq 1) \end{cases}$$

and

$$h(s, 1) = \begin{cases} f(2s) & \text{if } 1 \geq 4s - 1 \text{ (i.e., } 0 \leq s \leq \frac{1}{2}) \\ g(4s - 2) & \text{if } 4s - 1 \geq 1 \geq 4s - 2 \text{ (i.e., } \frac{1}{2} \leq s \leq \frac{3}{4}) \\ k(4s - 3) & \text{if } 4s - 2 \geq 1 \text{ (i.e., } \frac{3}{4} \leq s \leq 1). \end{cases}$$

Thus $h(s, 0) = [(f * g) * k](s)$ and $h(s, 1) = [f * (g * k)](s)$ for all $s \in I^1$. Also, $h(0, t) = f(0) = x_0 = k(1) = h(1, t)$ for all $t \in I^1$. Hence $(f * g) * k \;\tilde{\tilde{x}}_0\; f * (g * k)$.

(2) We show that the constant mapping $c : I^1 \to \{x_0\}$ is such that $[c]$ is the identity element of $\prod_1 (S, x_0)$ with respect to "$\circ$." Thus we must show that $f * c \;\tilde{\tilde{x}}_0\; f$ for any $f \in C(S, x_0)$. Let $h : I^2 \to S$ be defined as follows:

$$h(s, t) = \begin{cases} f\left(\dfrac{2s}{1 + t}\right) & \text{if } \langle s, t\rangle \in I^2 \text{ and } t \geq 2s - 1 \\ x_0 & \text{if } \langle s, t\rangle \in I^2 \text{ and } 2s - 1 \geq t. \end{cases}$$

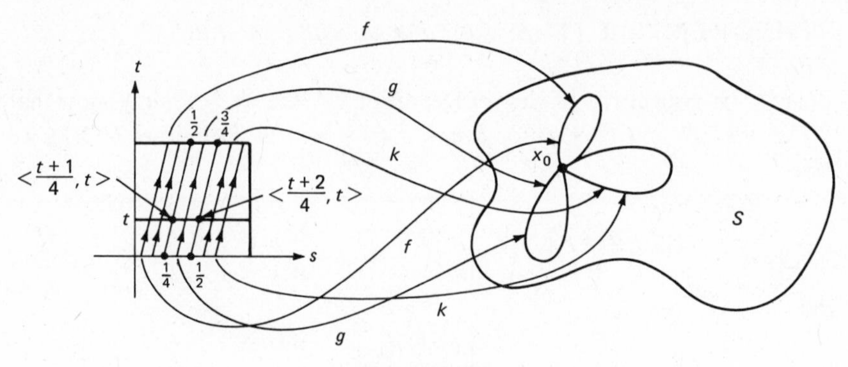

FIGURE 7-3

Figure 7-4 illustrates this situation. Since f is continuous, h is continuous, and we have

$$h(s, 0) = \begin{cases} f(2s) & \text{if } 0 \geq 2s - 1 \text{ (i.e., } 0 \leq s \leq \frac{1}{2}) \\ x_0 & \text{if } 2s - 1 \geq 0 \text{ (i.e., } \frac{1}{2} \leq s \leq 1) \end{cases}$$

and

$$h(s, 1) = f(s) \qquad \text{if } 1 \geq 2s - 1 \text{ (i.e., } 0 \leq s \leq 1).$$

Thus $h(s, 0) = (f * c)(s)$ and $h(s, 1) = f(s)$ for all $s \in I^1$. Moreover, we have $h(0, t) = f(0) = x_0 = f(1) = h(1, t)$ for all $t \in I^1$. Hence $f * c \;\tilde{\tilde{x}}_0\; f$.

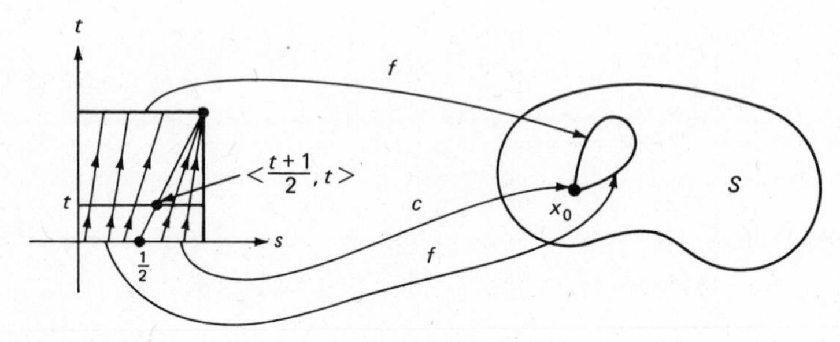

FIGURE 7-4

(3) Finally, we must show that each homotopy class $[f] \in \prod_1 (S, x_0)$ has an inverse $[g] \in \prod_1 (S, x_0)$ such that $[f] \circ [g] = [c]$. Thus we must show that if $f \in C(S, x_0)$, there exists $g \in C(S, x_0)$ such that $f * g \;\tilde{\tilde{x}}_0\; c$. Let $g(s) = f(1 - s)$ for all $s \in I^1$. Since $g(0) = f(1) = x_0 = f(0) = g(1)$, $g \in C(S, x_0)$. By Definition 7-13(1), we have

$$(f * g)(s) = \begin{cases} f(2s) & \text{if } 0 \leq s \leq \frac{1}{2} \\ g(2s - 1) = f(2 - 2s) & \text{if } \frac{1}{2} \leq s \leq 1. \end{cases}$$

We define a homotopy h between $f * g$ and c as follows:

$$h(s, t) = \begin{cases} x_0 & \text{if } 0 \leq s \leq \frac{t}{2} \\ f(2s - t) & \text{if } \frac{t}{2} \leq s \leq \frac{1}{2} \\ g(2s + t - 1) & \text{if } \frac{1}{2} \leq s \leq 1 - \frac{t}{2} \\ x_0 & \text{if } 1 - \frac{t}{2} \leq s \leq 1. \end{cases}$$

Figure 7-5 illustrates this situation. Since f and g are continuous, h is continuous, and we have

$$h(s, 0) = \begin{cases} f(2s) & \text{if } 0 \leq s \leq \frac{1}{2} \\ g(2s - 1) & \text{if } \frac{1}{2} \leq s \leq 1 \end{cases}$$

and

$$h(s, 1) = x_0 \qquad \text{if } 0 \leq s \leq 1.$$

Thus $h(s, 0) = (f * g)(s)$ and $h(s, 1) = c(s) = x_0$ for all $s \in I^1$. Moreover, we have $h(0, t) = x_0 = h(1, t)$ for all $t \in I^1$. Hence $f * g \ \tilde{\tilde{x}}_0 \ c$.

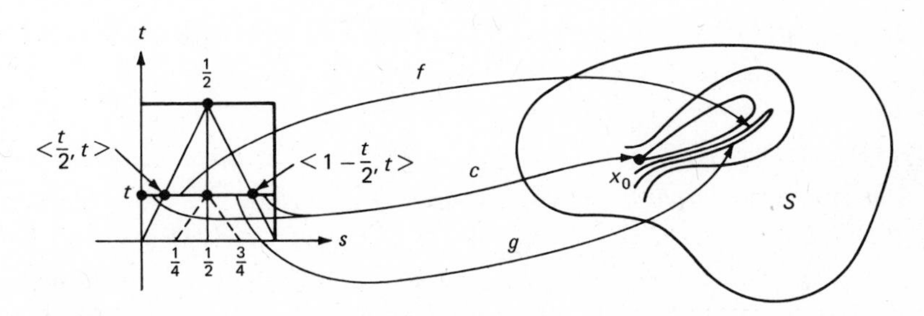

FIGURE 7-5

In general, $\prod_1 (S, x_0)$ depends upon x_0. However, in the case of an arcwise-connected space S, we can show that $\prod_1 (S, x_0)$ is independent of x_0. Thus we denote the fundamental group of an arcwise-connected space S simply by $\prod_1 (S)$. Also, $\prod_1 (S, x_0)$ is *not* Abelian in general, even if S is arcwise connected. However, if S is a special kind of arcwise-connected space (a "Hopf space"), then $\prod_1 (S)$ is Abelian. See Hocking and Young, *Topology*, pp. 167–69, for the details.

THEOREM 7-11. *If* S *is arcwise connected and* $x_0, x_1 \in S$, *then* $\prod_1 (S, x_0) \cong \prod_1 (S, x_1)$.

Proof. Since S is arcwise connected, there exists a homeomorphism $f: I^1 \to S$ with $f(0) = x_0$ and $f(1) = x_1$. Let $g(s) = f(1 - s)$ so that $g(0) = f(1) = x_1$ and $g(1) = f(0) = x_0$. Thus $f * g \;\tilde{\tilde{x}}_0\; c_0 : I^1 \to \{x_0\}$ and $g * f \;\tilde{\tilde{x}}_1\; c_1 : I^1 \to \{x_1\}$; i.e., $[f * g] = [c_0]$ and $[g * f] = [c_1]$, the identity elements in $\prod_1 (S, x_0)$ and $\prod_1 (S, x_1)$, respectively. If $[k] \in \prod_1 (S, x_0)$, let $\varphi([k]) = [g * k * f] \in \prod_1 (S, x_1)$. We observe that φ is a 1-1 mapping, since $k \;\tilde{\tilde{x}}_0\; l$ iff $g * k * f \;\tilde{\tilde{x}}_1\; g * l * f$ for $k, l \in C(S, x_0)$. Also, φ is onto, since $[k] \in \prod_1 (S, x_1)$ implies that $[k] = \varphi([f * k * g])$, where $f * k * g \in C(S, x_0)$. Finally, φ is an isomorphism, since $\varphi([k] \circ [l]) = \varphi([k]) \circ \varphi([l])$ for $[k]$, $[l] \in \prod_1 (S, x_0)$.

EXAMPLE 7-3. The fundamental group of the following spaces consists of just the identity: a contractible space, $S^n (n > 1)$, and $E^3 - \{p\}$.

EXAMPLE 7-4. The fundamental groups of the following spaces are infinite cyclic: S^1, $E^2 - \{p\}$, an annulus in E^2, and $E^3 - \{\text{a line } L\}$.

EXAMPLE 7-5. If T is the torus, then $\prod_1 (T) \cong Z \oplus Z$. If $S = E^2 - \{p_1, p_2\}$, then $\prod_1 (S)$ is a free group with two generators.

DEFINITION 7-14. $f: \langle S, A\rangle \to \langle T, B\rangle$ is a *pair mapping* iff $A \subset S$ and $B \subset T$ are closed, and $f: S \to T$ is a mapping with $f(A) \subset B$.

THEOREM 7-12. *If $f: \langle S, \{x_0\}\rangle \to \langle T, \{y_0\}\rangle$ is a pair mapping, then f induces a homomorphism $f_* : \prod_1 (S, x_0) \to \prod_1 (T, y_0)$.*

Proof. Let $f_1 : C(S, x_0) \to C(T, y_0)$ be given by $(f_1 l)(s) = f(l(s))$ for each $l \in C(S, x_0)$ and all $s \in I^1$. If $l_0 \in C(S, x_0)$ and $f_1 l_0 \in U$, a basic member of the compact-open topology on $C(T, y_0)$, then U consists of those closed paths on T based at y_0 which map a compact set $K \subset I^1$ into an open set $G \subset T$. Thus U^{-1} consists of those closed paths on S based at x_0 which map K into the open set $f^{-1}(G) \subset S$, and $l_0 \in U^{-1}$. If $g \in U^{-1}$, then $g(K) \subset f^{-1}(G)$ and $(f_1 g)(K) = f(g(K)) \subset G$. Thus $f_1(U^{-1}) \subset U$, so that f_1 is continuous and maps arcwise-connected components of $C(S, x_0)$ into arcwise-connected components of $C(T, y_0)$. Now let $f_*([l]) = [f_1 l]$ for $[l] \in \prod_1 (S, x_0)$. We must show that $f_*([l] \circ [g]) = f_*([l]) \circ f_*([g])$, which is true iff $f_1(l * g) = (f_1 l) * (f_1 g)$. We observe that

$$(f_1(l * g))(s) = \begin{cases} f(l(2s)) = (f_1 l)(2s) & \text{if } 0 \le s \le \frac{1}{2} \\ f(g(2s - 1)) = (f_1 g)(2s - 1) & \text{if } \frac{1}{2} \le s \le 1. \end{cases}$$

Thus $f_1(l * g) = (f_1 l) * (f_1 g)$, and f_* is a homomorphism.

THEOREM 7-13. *If* $f \simeq g: \langle S, \{x_0\}\rangle \to \langle T, \{y_0\}\rangle$, *then* $f_* = g_*:$ $\prod_1 (S, x_0) \to \prod_1 (T, y_0)$.

Proof. If $k \in C(S, x_0)$, then $f(k) \tilde{\bar{y}}_0 g(k)$. Hence $f_1 k \tilde{\bar{y}}_0 g_1 k$, and $f_* = g_*$.

THEOREM 7-14. *If* $f: \langle S, \{x_0\}\rangle \to \langle T, \{y_0\}\rangle$ *and* $g: \langle T, \{y_0\}\rangle \to \langle V, \{z_0\}\rangle$, *then* $(gf)_* = g_* f_*$.

Proof. If $k \in C(S, x_0)$, then $((gf)_1 k)(s) = (gf)(k(s)) = g(f(k(s))) = g_1(f(k(s))) = g_1(f_1(k(s))) = (g_1 f_1)(k(s))$. Thus $(gf)_1 = g_1 f_1$, which implies that $(gf)_* = g_* f_*$.

COROLLARY 7-3. *If* $\langle S, \{x_0\}\rangle$ *and* $\langle T, \{y_0\}\rangle$ *are homotopically equivalent, then* $\prod_1 (S, x_0) \cong \prod_1 (T, y_0)$.

Proof. Since $\langle S, \{x_0\}\rangle$ and $\langle T, \{y_0\}\rangle$ are homotopically equivalent, there exist mappings $f: \langle S, \{x_0\}\rangle \to \langle T, \{y_0\}\rangle$ and $g: \langle T, \{y_0\}\rangle \to \langle S, \{x_0\}\rangle$ such that $gf \simeq i_S$ and $fg \simeq i_T$. Thus $(fg)_* = f_* g_*$ and $(gf)_* = g_* f_*$ are isomorphisms onto. This implies that f_* and g_* are isomorphisms onto.

COROLLARY 7-4. *Any two homeomorphic spaces have isomorphic fundamental groups.*

Proof. Let $h: S \to T$ be a homeomorphism onto and $x_0 \in S$. Thus $\langle S, \{x_0\}\rangle$ and $\langle T, \{h(x_0)\}\rangle$ are homotopically equivalent. Hence $\prod_1 (S, x_0) \cong \prod_1 (T, h(x_0))$ by Corollary 7-3.

EXERCISES

7-10. Show that if $f_1 \tilde{\bar{x}}_0 f_2$ and $g_1 \tilde{\bar{x}}_0 g_2$, then $f_1 * g_1 \tilde{\bar{x}}_0 f_2 * g_2$ for $f_1, f_2, g_1, g_2 \in C(S, x_0)$.

7-11. Supply the missing details to show that the function φ defined in the proof of Theorem 7-11 is a 1-1 homomorphism onto (and hence an isomorphism).

7-12. What is the fundamental group of the closed n-cell B^n?

★7-6 The Higher Homotopy Groups

In this final section on homotopy, we define higher homotopy groups $\prod_n (S, x_0)$, $n > 1$, and point out that they are all Abelian.

DEFINITION 7-15. If $f: I^n \to S$ is continuous, then f is a *closed path on S based at* $x_0 \in S$ iff f maps the boundary βI^n of I^n onto $\{x_0\}$. For each

$x_0 \in S$, denote by $C_n(S, x_0)$ the collection of all closed paths on S based at x_0 with the compact-open topology.

DEFINITION 7-16. If $f, g \in C_n(S, x_0)$, then f and g are *homotopic modulo* x_0 ($f \tilde{\tilde{x}}_0\, g$) iff there exists a continuous mapping $h : I^{n+1} \to S$ such that $h(\bar{s}, 0) = f(\bar{s})$ and $h(\bar{s}, 1) = g(\bar{s})$ for all $\bar{s} = \langle s_1, \ldots, s_n \rangle \in I^n$ and $h(\beta I^n, t) = x_0$ for all $t \in I^1$.

As before, homotopy modulo x_0 is an equivalence relation on $C_n(S, x_0)$, and the resulting collection of equivalence (homotopy) classes is denoted by $\prod_n (S, x_0)$.

DEFINITION 7-17

(1) If $f, g \in C_n(S, x_0)$, the *juxtaposition* $f * g$ of f and g is defined as follows:

$$(f * g)(s_1, \ldots, s_n) = \begin{cases} f(2s_1, s_2, \ldots, s_n) & \text{if } 0 \leq s_1 \leq \frac{1}{2} \\ g(2s_1 - 1, s_2, \ldots, s_n) & \text{if } \frac{1}{2} \leq s_1 \leq 1. \end{cases}$$

(2) If $[f], [g] \in \prod_n (S, x_0)$, we define $[f] \circ [g] = [f * g]$, as before.

THEOREM 7-15. $\prod_n (S, x_0)$ *is an Abelian group with respect to* "$\circ$" *for* $n > 1$.

EXERCISES

7-13. Show that homotopy modulo x_0 is an equivalence relation on $C_n(S, x_0)$.

7-14. Show that if $f_1 \tilde{\tilde{x}}_0 f_2$ and $g_1 \tilde{\tilde{x}}_0 g_2$, then $f_1 * g_1 \tilde{\tilde{x}}_0 f_2 * g_2$ for $f_1, f_2, g_1, g_2 \in C_n(S, x_0)$.

CHAPTER 8
Singular Homology Theory

8-1 Simplexes and Complexes

In this chapter the reader is introduced to homology theory. We define the integral (singular) homology groups associated with a topological pair $\langle S, A\rangle$, where S is a topological space and A is a subspace of S. In case $A = \emptyset$, we shall speak about the homology groups of S rather than of $\langle S, \emptyset\rangle$. We also discuss those properties of singular homology theory which constitute the classical "Eilenberg–Steenrod axioms" for a homology theory. We define the singular homology sequence for a topological pair $\langle S, A\rangle$ and show that it is exact. Then, by way of application of these ideas, we present some sample computations of homology groups and prove several well-known theorems, including the Brouwer fixed-point theorem and the Fundamental Theorem of Algebra. Since this is intended to be only an introduction to homology theory, we develop only the theoretically simpler singular homology theory, rather than the more intuitive simplicial theory. Also, the topics of Čech homology and cohomology are omitted, since they really belong in a graduate-level treatment of algebraic topology.

In this section we introduce the notions of singular simplexes and complexes. These concepts are then used in defining the notions of singular chains, cycles, and boundaries. This discussion is a prelude to our defining in Section 8-3 the integral (singular) homology groups of a topological pair $\langle S, A\rangle$.

DEFINITION 8-1. The *standard n-simplex* is the set

$$\begin{aligned}\Delta_n &= [e_0 e_1 \cdots e_n]\\ &= \Big\{\langle x_0, x_1, \ldots, x_n\rangle \in E^{n+1} \mid x_i \geq 0\\ &\qquad\qquad (i = 0, 1, \ldots, n) \text{ and } \sum_{i=0}^{n} x_i = 1\Big\}.\end{aligned}$$

The point $e_{i-1} \in \Delta_n$ having its ith coordinate $x_{i-1} = 1$ and the rest of its coordinates equal to zero is called the ith vertex of Δ_n $(i = 1, 2, \ldots, n + 1)$. The subspace $\Delta_n^{(i)} = \{\langle x_0, \ldots, x_i, \ldots, x_n\rangle \in \Delta_n \mid x_i = 0\}$ is the $(n - 1)$-*face* of Δ_n opposite the vertex e_i $(i = 0, 1, \ldots, n)$.

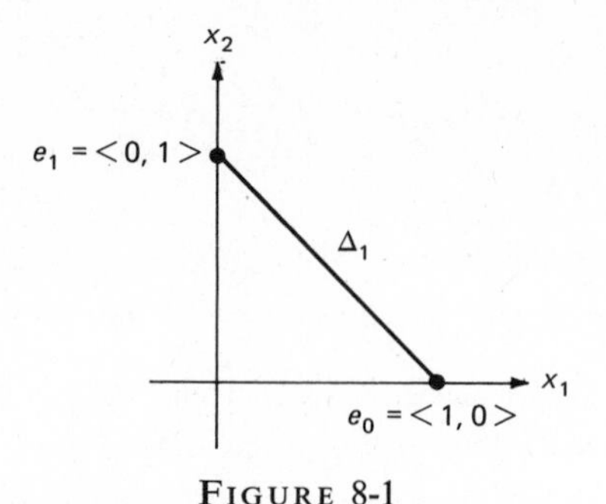

FIGURE 8-1

EXAMPLE 8-1. The standard 1-simplex Δ_1 has two 0-faces ($\{e_0\}$ and $\{e_1\}$), as shown in Figure 8-1. The standard 2-simplex Δ_2 has the three 1-faces ($[e_0e_1]$, $[e_1e_2]$, and $[e_0e_2]$) and three 0-faces ($\{e_0\}$, $\{e_1\}$, and $\{e_2\}$), as shown in Figure 8-2.

If $n \in I^+$, we define for each $i = 0, 1, \ldots, n$ a homeomorphism $v_i : \Delta_{n-1} \to \Delta_n^{(i)}$ by the rule

$$v_i(\langle x_0, \ldots, x_{n-1}\rangle) = \langle x_0, \ldots, x_{i-1}, 0, x_i, \ldots, x_{n-1}\rangle$$

for each $\langle x_0, \ldots, x_{n-1}\rangle \in \Delta_{n-1}$. Then if $v_i : \Delta_{n-2} \to \Delta_{n-1}$ and $v_j : \Delta_{n-1} \to \Delta_n$ are two such homeomorphisms with $n \in I^+$ and $0 \leq i < j \leq n$, it follows that $v_j \circ v_i = v_i \circ v_{j-1}$ is a homeomorphism of Δ_{n-2} into Δ_n. Details are left to the reader as an exercise.

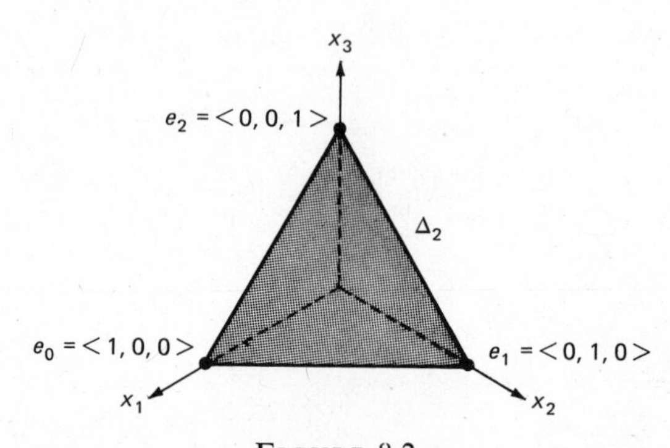

FIGURE 8-2

We are now ready to define the concept "singular n-simplex," which is basic to the development of singular homology theory. We remind the reader at this point that the word "mapping" means for us a continuous function and is used in this way throughout this chapter.

DEFINITION 8-2. If S is a topological space, a *singular n-simplex* in S is a mapping $\sigma : \Delta_n \to S$. The modifier "singular" refers to the fact that σ need not be a homeomorphism but merely continuous.

DEFINITION 8-3. If S is a topological space and $n \geq 0$ is an integer, then $\mathcal{S}_n(S)$ denotes the collection of all singular n-simplexes in S. Moreover, the *singular complex* of S is the set $\mathcal{S}(S) = \bigcup\{\mathcal{S}_n(S) \mid n \geq 0\}$ with the face operators described in Definition 8-4.

EXAMPLE 8-2. Let S be any topological space. Since the mapping f given by the rule $f(t) = \langle 1 - t, t\rangle$ for each $t \in I^1$ is a homeomorphism of I^1 onto Δ_1, any singular 1-simplex σ in S can be regarded as a path $\sigma \circ f : I^1 \to S$ from $\sigma(e_0)$ to $\sigma(e_1)$ in S. Thus the collection $\mathcal{S}_1(S)$ of all singular 1-simplexes in S is the same as the collection of all paths in S.

DEFINITION 8-4. If $n \in I^+$ and $\sigma \in \mathcal{S}_n(S)$, then the composition $\sigma \circ v_i : \Delta_{n-1} \to S$ is a singular $(n - 1)$-simplex in S which we call the ith *face* of σ and denote by $\sigma^{(i)}$ $(i = 0, 1, \ldots, n)$. This determines for each $i = 0, 1, \ldots, n$, a function $d_i : \mathcal{S}_n(S) \to \mathcal{S}_{n-1}(S)$ given by $d_i(\sigma) = \sigma^{(i)}$ for each $\sigma \in \mathcal{S}_n(S)$, which we call the ith-*face operator* on $\mathcal{S}(S)$.

DEFINITION 8-5. If A is a subspace of a topological space S, and if $\sigma \in \mathcal{S}_n(A)$, then σ may be regarded as the singular n-simplex $i \circ \sigma : \Delta_n \to S$, where i is the inclusion mapping of A into S. Thus we have that $\mathcal{S}_n(A) \subset \mathcal{S}_n(S)$ and $\mathcal{S}(A) \subset \mathcal{S}(S)$; i.e., $\mathcal{S}(A)$ is a *subcomplex* of $\mathcal{S}(S)$.

EXERCISES

8-1. If S is any topological space, describe the collection $\mathcal{S}_0(S)$ of singular 0-simplexes in S.

8-2. Verify that the mapping $v_i : \Delta_{n-1} \to \Delta_n^{(i)}$, defined in this section, is a homeomorphism onto for each integer $i = 0, 1, \ldots, n$.

8-3. Supply the details necessary to show that $v_j \circ v_i = v_i \circ v_{j-1}$, where $v_i : \Delta_{n-2} \to \Delta_{n-1}$ and $v_j : \Delta_{n-1} \to \Delta_n$ are the two homeomorphisms described in this section $(0 \leq i < j \leq n)$.

8-4. If $\sigma \in \mathcal{S}_n(S)$, where $n > 1$, and if $0 \leq i < j \leq n$, show that $[\sigma^{(j)}]^{(i)} = [\sigma^{(i)}]^{(j-1)}$. (*Hint:* Use Exercise 8-3.)

8-2 Singular Chains

In this section we define the n-dimensional singular chain-group $\mathscr{C}_n(S, A)$ of a topological pair $\langle S, A\rangle$ for each integer n. Next, we define the boundary operator $d: \mathscr{C}_n(S, A) \to \mathscr{C}_{n-1}(S, A)$, which is a homomorphism for each $n \in I^+$. Whenever necessary to avoid confusion, we shall use the notation d^m for the boundary operator on $\mathscr{C}_m(S, A)$ and d^n for the boundary operator on $\mathscr{C}_n(S, A)$ when both are being considered in the same discussion. Using this terminology, we define the sequence

$$\cdots \longrightarrow \mathscr{C}_{n+1}(S, A) \xrightarrow{d^{n+1}} \mathscr{C}_n(S, A) \xrightarrow{d^n} \mathscr{C}_{n-1}(S, A) \longrightarrow \cdots$$

to be integral singular chain complex of $\langle S, A\rangle$. Also for each integer n, we define the chain-group homomorphism $f_n: \mathscr{C}_n(S, A) \to \mathscr{C}_n(T, B)$ to be that which is induced by a pair mapping $f: \langle S, A\rangle \to \langle T, B\rangle$. As a first stage in this development, we consider the case where $A = \varnothing$. In this case we can replace $\langle S, \varnothing\rangle$ by simply S, since $\mathscr{S}_n(\varnothing) = \varnothing$ implies that $\mathscr{C}_n(S, \varnothing) = \mathscr{C}_n(S)$.

DEFINITION 8-6. For each integer $n \geq 0$, the *n-dimensional singular chain-group* of S is the additive Abelian group $\mathscr{C}_n(S)$ generated by the set $\mathscr{S}_n(S)$ of singular n-simplexes in S, using the additive group Z of integers as our coefficient group. Thus the elements of $\mathscr{C}_n(S)$ (i.e., the *singular n-chains*) are of the form

$$\alpha = m_1\sigma_1 + m_2\sigma_2 + \cdots + m_k\sigma_k,$$

where $m_i \in Z$, $\sigma_i \in \mathscr{S}_n(S)$ $(i = 1, 2, \ldots, k)$ and $k \in I^+$. For each integer $n < 0$, let $\mathscr{C}_n(S) = \{0\}$ be the trivial group consisting of just the additive identity.

DEFINITION 8-7. For each integer $n > 0$, the *boundary operator* $d: \mathscr{C}_n(S) \to \mathscr{C}_{n-1}(S)$ is the homomorphism defined as follows:

(1) If $\sigma \in \mathscr{S}_n(S)$, then $d\sigma = \sum_{i=0}^{n} (-1)^i d_i\sigma = \sum_{i=0}^{n} (-1)^i \sigma^{(i)} \in \mathscr{C}_{n-1}(S)$.

(2) If $\alpha = \sum_{i=1}^{k} m_i\sigma_i \in \mathscr{C}_n(S)$, then $d\alpha = \sum_{i=1}^{k} m_i d\sigma_i \in \mathscr{C}_{n-1}(S)$ and is called the boundary of α.

For each integer $n < 0$, we let $d: \mathscr{C}_n(S) \to \mathscr{C}_{n-1}(S)$ be the trivial homomorphism, since $\mathscr{C}_{n-1}(S) = \{0\}$ in that case.

From the above definition, it is clear that the boundary of an n-dimensional chain on S is an $(n - 1)$-dimensional chain on S. If the boundary operator is applied a second time, the result is the additive identity "0" of $\mathscr{C}_{n-2}(S)$; i.e., the composition of the boundary operator with itself is the trivial homomorphism.

THEOREM 8-1. *If* $d^n : \mathscr{C}_n(S) \to \mathscr{C}_{n-1}(S)$ *and* $d^{n-1} : \mathscr{C}_{n-1}(S) \to \mathscr{C}_{n-2}(S)$ *are boundary operators, then* $d^{n-1} \circ d^n : \mathscr{C}_n(S) \to \mathscr{C}_{n-2}(S)$ *is the trivial homomorphism* [*i.e., the image of* $\mathscr{C}_n(S)$ *under* $d^{n-1} \circ d^n$ *is* $\{0\}$] *for all n.*

Proof. We assume that $n \geq 2$, since otherwise $\mathscr{C}_{n-2}(S) = \{0\}$, and the result is clear. If $\sigma \in \mathscr{S}_n(S)$, then

$$d^{n-1} \circ d^n(\sigma) = d^{n-1}[d^n\sigma] = d^{n-1}\left[\sum_{i=0}^{n} (-1)^i \sigma^{(i)}\right] \qquad \text{(Definition 8-7)}$$

$$= \sum_{i=0}^{n} (-1)^i d^{n-1}\sigma^{(i)} = \sum_{i=0}^{n} (-1)^i \sum_{j=0}^{n-1} (-1)^j [\sigma^{(i)}]^{(j)}$$

(Definition 8-7)

$$= \sum_{0 \leq j < i \leq n} (-1)^{i+j}[\sigma^{(i)}]^{(j)} + \sum_{0 \leq i \leq j < n} (-1)^{i+j}[\sigma^{(i)}]^{(j)}$$

$$= \sum_{0 \leq j < i \leq n} (-1)^{i+j}[\sigma^{(j)}]^{(i-1)} + \sum_{0 \leq i \leq j < n} (-1(^{i+j}[\sigma^{(i)}]^{(j)}$$

(Exercise 8-4)

$$= (-1) \sum_{0 \leq i \leq j < n} (-1)^{i+j}[\sigma^{(i)}]^{(j)} + \sum_{0 \leq i \leq j < n} (-1)^{i+j}[\sigma^{(i)}]^{(j)}$$

$$= 0$$

as the result of replacing j by i and $i - 1$ by j in the first summation. If $\alpha = \sum_{i=1}^{k} m_i\sigma_i \in \mathscr{C}_n(S)$, then $d^{n-1} \circ d^n(\alpha) = \sum_{i=1}^{k} m_i d^{n-1} \circ d^n(\sigma_i) = 0$, since we showed that $d^{n-1} \circ d^n(\sigma_i) = 0$ $(i = 1, \ldots, k)$.

DEFINITION 8-8. For any topological space S, the sequence

$$\cdots \longrightarrow \mathscr{C}_{n+1}(S) \xrightarrow{d^{n+1}} \mathscr{C}_n(S) \xrightarrow{d^n} \mathscr{C}_{n-1}(S) \longrightarrow \cdots$$

is called the *integral singular chain complex* of S.

Any mapping of a topological space S into a topological space T induces a homomorphism of $\mathscr{C}_n(S)$ into $\mathscr{C}_n(T)$ for each integer n as follows.

DEFINITION 8-9. Let S and T be topological spaces, and let $f: S \to T$ be continuous. If $\sigma \in \mathscr{S}_n(S)$, then $f_n(\sigma) = f \circ \sigma \in \mathscr{S}_n(T)$. Moreover, for $\alpha = \sum_{i=1}^{k} m_i\sigma_i \in \mathscr{C}_n(S)$, we let $f_n(\alpha) = \sum_{i=1}^{k} m_i f_n(\sigma_i) \in \mathscr{C}_n(T)$. For each integer n, the mapping $f_n : \mathscr{C}_n(S) \to \mathscr{C}_n(T)$ defined in this manner is the *chain-group homomorphism* induced by f.

REMARK. In a natural way we can extend the above discussion to the general case of a topological pair $\langle S, A \rangle$, where A is any subspace of S.

(1) For each integer $n \geq 0$, the *n-dimensional singular chain group* $\mathscr{C}_n(S, A)$ of the topological pair $\langle S, A \rangle$ is the quotient group $\mathscr{C}_n(S)/\mathscr{C}_n(A)$, where $\mathscr{C}_n(A)$ is generated by $\mathscr{S}_n(A)$. The elements of $\mathscr{C}_n(S, A)$ are called the *singular n-chains of* $\langle S, A \rangle$ or the *singular n-chains of S modulo A*. $\mathscr{C}_n(S, A)$ is isomorphic to that subgroup of $\mathscr{C}_n(S)$ which is generated by $\mathscr{S}_n(S) - \mathscr{S}_n(A)$. For each integer $n < 0$, we let $\mathscr{C}_n(S, A) = \{0\}$ be the trivial subgroup of $\mathscr{C}_n(S)$ containing just the additive identity.

(2) If $d : \mathscr{C}_n(S) \to \mathscr{C}_{n-1}(S)$ is the boundary operator, then the restriction $d \mid \mathscr{C}_n(A)$ maps $\mathscr{C}_n(A)$ into $\mathscr{C}_{n-1}(A)$. Thus d induces a homomorphism of the quotient group $\mathscr{C}_n(S, A)$ into the quotient group $\mathscr{C}_{n-1}(S, A)$, which we denote also by d or d^n when necessary to avoid confusion between boundary operators on chain groups of different dimensions n. The reader should verify that Theorem 8-1 is also valid for this induced boundary homomorphism $d : \mathscr{C}_n(S, A) \to \mathscr{C}_{n-1}(S, A)$.

(3) The sequence

$$\cdots \longrightarrow \mathscr{C}_{n+1}(S, A) \xrightarrow{d^{n+1}} \mathscr{C}_n(S, A) \xrightarrow{d^n} \mathscr{C}_{n-1}(S, A) \longrightarrow \cdots$$

is called the *integral singular chain complex* of $\langle S, A \rangle$.

(4) If $f : \langle S, A \rangle \to \langle T, B \rangle$ is a pair mapping, then f induces a homomorphism $f_n : \mathscr{C}_n(S, A) \to \mathscr{C}_n(T, B)$ called the *chain-group homomorphism* for each integer n. If $\sigma \in \mathscr{S}_n(S)$, then $f_n(\sigma) = f \circ \sigma \in \mathscr{S}_n(T)$, and if $\sigma \in \mathscr{S}_n(A)$, then $f_n(\sigma) = f \circ \sigma \in \mathscr{S}_n(B)$. Clearly, f_n extends to a homomorphism of $\mathscr{C}_n(S)$ into $\mathscr{C}_n(T)$, which takes $\mathscr{C}_n(A)$ into $\mathscr{C}_n(B)$. The required homomorphism is this pair homomorphism $f_n : \langle \mathscr{C}_n(S), \mathscr{C}_n(A) \rangle \to \langle \mathscr{C}_n(T), \mathscr{C}_n(B) \rangle$.

EXERCISES

8-5. Show that $\mathscr{C}_n(S, \varnothing) = \mathscr{C}_n(S)$ for each integer n.

8-6. Show that if $d^n : \mathscr{C}_n(S, A) \to \mathscr{C}_{n-1}(S, A)$ and $d^{n-1} : \mathscr{C}_{n-1}(S, A) \to \mathscr{C}_{n-2}(S, A)$ are the restricted boundary operators, then $d^{n-1} \circ d^n : \mathscr{C}_n(S, A) \to \mathscr{C}_{n-2}(S, A)$ is the trivial homomorphism.

8-7. Let $f : \langle S, A \rangle \to \langle T, B \rangle$ and $g : \langle T, B \rangle \to \langle W, C \rangle$ be pair mappings. If for each integer $n \geq 0$, $f_n : \mathscr{C}_n(S, A) \to \mathscr{C}_n(T, B)$, $g_n : \mathscr{C}_n(T, B) \to \mathscr{C}_n(W, C)$, and $(g \circ f)_n : \mathscr{C}_n(S, A) \to \mathscr{C}_n(W, C)$ are the respective induced chain-group homomorphisms, show that $(g \circ f)_n = g_n \circ f_n$.

8-8. Let $f: \langle S, A \rangle \to \langle T, B \rangle$ be a pair mapping. Prove that the boundary operator commutes with the induced chain-group homomorphism; i.e., show that $f_n \circ d = d \circ f_{n+1}$.

8-3 Singular Homology Groups

The n-dimensional singular chain group $\mathscr{C}_n(S, A)$ of the topological pair $\langle S, A \rangle$ contains two important subgroups. These are the group $\mathscr{Z}_n(S, A)$ of singular "n-cycles" in S modulo A and the group $\mathscr{B}_n(S, A)$ of singular "n-boundaries" in S modulo A. These are described in the discussion below and used to define the singular homology group $H_n(S, A)$ for the pair $\langle S, A \rangle$, for each integer n.

DEFINITION 8-10. If $\alpha \in \mathscr{C}_n(S, A)$, then α is an *n-cycle* in S modulo A iff $d\alpha = 0$. Thus the collection $\mathscr{Z}_n(S, A)$ of n-cycles in S modulo A is just the kernel of the boundary operator $d: \mathscr{C}_n(S, A) \to \mathscr{C}_{n-1}(S, A)$ and a subgroup of $\mathscr{C}_n(S, A)$.

DEFINITION 8-11. If $\alpha \in \mathscr{C}_n(S, A)$, then α is an *n-boundary* in S modulo A iff there exists $\gamma \in \mathscr{C}_{n+1}(S, A)$ such that $\alpha = d\gamma$. Thus the collection $\mathscr{B}_n(S, A)$ of n-boundaries in S modulo A is the image of the boundary operator $d: \mathscr{C}_{n+1}(S, A) \to \mathscr{C}_n(S, A)$ and is a subgroup of $\mathscr{C}_n(S, A)$.

Using Exercise 8-6, the reader may easily verify that $\mathscr{B}_n(S, A)$ is actually a subgroup of $\mathscr{Z}_n(S, A)$. We now define for each integer n the singular homology group $H_n(S, A)$ to be the quotient group $\mathscr{Z}_n(S, A)/\mathscr{B}_n(S, A)$.

DEFINITION 8-12. The *n-dimensional singular homology group* of the pair $\langle S, A \rangle$ is the quotient group $H_n(S, A) = \mathscr{Z}_n(S, A)/\mathscr{B}_n(S, A)$. The elements of $H_n(S, A)$ are *homology* classes and are the cosets $[\alpha] = \alpha + \mathscr{B}_n(S, A)$, where $\alpha \in \mathscr{Z}_n(S, A)$. Thus, if $\alpha, \beta \in \mathscr{Z}_n(S, A)$, then α and β are *homologous modulo A* or belong to the same homology class ($\alpha \sim \beta \bmod A$) iff $\alpha - \beta \in \mathscr{B}_n(S, A)$.

REMARK. In the special case in which $A = \varnothing$, we have $\mathscr{C}_n(S, \varnothing) = \mathscr{C}_n(S)$, $\mathscr{Z}_n(S, \varnothing) = \mathscr{Z}_n(S)$, $\mathscr{B}_n(S, \varnothing) = \mathscr{B}_n(S)$, and $H_n(S, \varnothing) = H_n(S) = \mathscr{Z}_n(S)/\mathscr{B}_n(S)$.

Since any singular 0-simplex is a constant mapping of Δ_0 into a point $x_0 \in S$ and thus can be identified with x_0, we have $\mathscr{S}_0(S) = S$. Thus the 0-dimensional chain group on S is

$$\mathscr{C}_0(S) = \left\{ \sum_{i=1}^{k} m_i x_i \mid m_i \in Z,\ x_i \in S, \text{ and } k \in I^+ \right\}.$$

Also, $\mathscr{C}_{-1}(S) = \{0\}$, which implies that $\mathscr{Z}_0(S)$ (= kernel of $d: \mathscr{C}_0(S) \to \mathscr{C}_{-1}(S)) = \mathscr{C}_0(S)$. If $\mathscr{P}(S)$ denotes the additive Abelian group generated by the collection of path components of S, it can be shown that there exists a unique homomorphism

$$h: \mathscr{Z}_0(S) \xrightarrow{\text{onto}} \mathscr{P}(S),$$

since $\mathscr{Z}_0(S) = \mathscr{C}_0(S)$ and $\mathscr{C}_0(S)$ is the additive Abelian group generated by S. Moreover, the kernel of h can be shown to be $\mathscr{B}_0(S)$. Thus, by the Fundamental Isomorphism Theorem (Theorem 0-5), there exists an isomorphism of $H_0(S) = \mathscr{Z}_0(S)/\mathscr{B}_0(S)$ onto $\mathscr{P}(S)$. This is an outline of the proof of Theorem 8-2, which has as an important consequence the fact that $H_0(S)$ is infinite cyclic iff S is pathwise connected. Details may be found in Hu's *Homology Theory*, pp. 214–18.

THEOREM 8-2. *$H_0(S)$ is isomorphic to the additive Abelian group $\mathscr{P}(S)$ generated by the collection of path components of S.*

EXERCISES

8-9. Verify that $\mathscr{B}_n(S, A)$ is a subgroup of $\mathscr{Z}_n(S, A)$. Moreover, if $\alpha \in \mathscr{C}_n(S)$, show that (a) $\alpha \in \mathscr{Z}_n(S, A)$ iff $d\alpha \in \mathscr{C}_{n-1}(A)$ and (b) $\alpha \in \mathscr{B}_n(S, A)$ iff $\exists \gamma \in \mathscr{C}_{n+1}(S, A)$ and $\exists \beta \in \mathscr{C}_n(A)$ such that $\alpha = d\gamma + \beta$.

8-10. Determine $H_n(S, A)$ for each integer $n < 0$.

8-11. Formulate the analogue of Theorem 8-2 for the topological pair $\langle S, A \rangle$ and outline its proof.

8-12. Use Theorem 8-2 and Exercise 8-11 to show that $H_0(S)$ is infinite cyclic and $H_0(S, A)$ is trivial in case S is pathwise connected and $\varnothing \neq A \subset S$.

8-13. If S is a singleton, show that $H_0(S)$ is infinite cyclic and $H_n(S)$ is trivial if $n \neq 0$.

8-14. If S is the discrete space $\{x_1, x_2, \ldots, x_n\}$, show that $H_0(S)$ is free Abelian with n generators and $H_m(S)$ is trivial if $m \neq 0$.

8-15. Prove that the relation "α and β are homologous n-cycles in S modulo A" is an equivalence relation on $\mathscr{Z}_n(S, A)$. The resulting equivalence classes (cosets) are the homology classes described in Definition 8-12.

8-4 Properties of Singular Homology

In this section we prove that a pair mapping $f: \langle S, A\rangle \to \langle T, B\rangle$ induces a homomorphism f_* of $H_n(S, A)$ into $H_n(T, B)$ by showing that the induced chain-group homomorphism

$$f_n : \mathscr{C}_n(S, A) \to \mathscr{C}_n(T, B)$$

maps homologous n-cycles in S modulo A onto homologous n-cycles in T modulo B. We show that the homology groups are topological invariants and that homotopically equivalent spaces have isomorphic homology groups. Consequently, homology is a weaker topological property than homotopy, since there exist spaces having the same homology groups which are not homotopically equivalent [e.g., H_n (torus) $\cong H_n$ ("punctured" torus), $\forall n \neq 2$, but the torus and "punctured" torus are not homotopically equivalent]. We conclude the section with a statement of the "Excision Theorem," which will be used later in our actual calculation of singular homology groups (Sections 8-5 and 8-6).

THEOREM 8-3. *If $f: \langle S, A\rangle \to \langle T, B\rangle$ is a pair mapping, then*
(1) $f_n(\mathscr{Z}_n(S, A)) \subset \mathscr{Z}_n(T, B)$.
(2) $f_n(\mathscr{B}_n(S, A)) \subset \mathscr{B}_n(T, B)$.
(3) $f_* : H_n(S, A) \to H_n(T, B)$ *is a homomorphism, where* $f_*([\alpha]) = [f_n(\alpha)]$ *for each* $[\alpha] \in H_n(S, A)$.

Proof

(1) If $\alpha \in \mathscr{Z}_n(S, A)$, then $d\alpha = 0$, which implies that $df_n(\alpha) = f_{n-1}(d\alpha) = f_{n-1}(0) = 0$, since d commutes with f_n, and f_n is a homomorphism. Thus $f_n(\alpha) \in \mathscr{Z}_n(T, B)$.

(2) If $\alpha \in \mathscr{B}_n(S, A)$, then $\exists \gamma \in \mathscr{C}_{n+1}(S, A)$ such that $\alpha = d\gamma$. This implies that $f_n(\alpha) = f_n(d\gamma) = df_{n+1}(\gamma)$, where $f_{n+1}(\gamma) \in \mathscr{C}_{n+1}(T, B)$. Thus $f_n(\alpha) \in \mathscr{B}_n(T, B)$.

(3) If $\alpha' \in [\alpha] \in H_n(S, A)$, then $\alpha - \alpha' \in \mathscr{B}_n(S, A)$. Hence $f_n(\alpha - \alpha') = f_n(\alpha) - f_n(\alpha') \in \mathscr{B}_n(T, B)$ by (2). Thus $f_*([\alpha]) = [f_n(\alpha)] = [f_n(\alpha')] = f_*([\alpha'])$, which implies that f_* is well defined (i.e., independent of the representative chosen from any homology class). Finally, $f_*([\alpha] + [\beta]) = f_*([\alpha + \beta]) =$

$[f_n(\alpha + \beta)] = [f_n(\alpha) + f_n(\beta)] = [f_n(\alpha)] + [f_n(\beta)] = f_*([\alpha]) + f_*([\beta])$, which implies that f_* is a homomorphism.

THEOREM 8-4. *If* $f: \langle S, A\rangle \to \langle T, B\rangle$ *and* $g: \langle T, B\rangle \to \langle W, C\rangle$ *are pair mappings, then* $(g \circ f)_* = g_* \circ f_*$.

Proof. If $[\alpha] \in H_n(S, A)$, then $(g \circ f)_*([\alpha]) = [(g \circ f)_n(\alpha)] = [g_n \circ f_n(\alpha)] = g_*([f_n(\alpha)]) = g_* \circ f_*([\alpha])$.

Using Theorem 8-4 with $\langle W, C\rangle = \langle S, A\rangle$ and $g = f^{-1}$, we are able to show that the homology groups of $\langle S, A\rangle$ are topological invariants.

THEOREM 8-5. *The singular homology groups of* $\langle S, A\rangle$ *are topological invariants.*

Proof. Let $f: S \xrightarrow{\text{onto}} T$ be a homeomorphism, and let $A \subset S$. Then $f: \langle S, A\rangle \to \langle T, f(A)\rangle$ and $g = f^{-1}: \langle T, f(A)\rangle \to \langle S, A\rangle$ are pair mappings onto; i.e., $g \circ f$ is the identity mapping on $\langle S, A\rangle$ and $f \circ g$ is the identity mapping on $\langle T, f(A)\rangle$. Hence $(g \circ f)_* = g_* \circ f_*$ is the identity mapping on $H_n(S, A)$ and $(f \circ g)_* = f_* \circ g_*$ is the identity mapping on $H_n(T, f(A))$. Thus $H_n(S, A) \cong H_n(T, f(A))$.

It seems reasonable to expect that two homotopic pair mappings should induce the same homomorphism of the singular homology groups. This result is known in homology theory as the "Homotopy Theorem." To prove it, we first introduce the notion of a "lifting map" and state as a lemma a fundamental property of lifting maps. A proof of this lemma can be found in Wallace's *An Introduction to Algebraic Topology*, pp. 128–29.

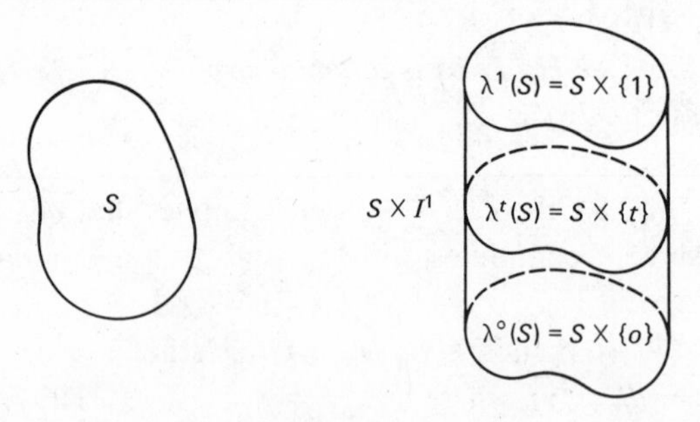

FIGURE 8-3

DEFINITION 8-13. Let S be a topological space and $\lambda^t: S \to S \times I^1$ be a mapping defined by $\lambda^t(x) = \langle x, t\rangle$ for each $x \in S$. For each $t \in I^1$, the mapping λ^t is called a *lifting mapping*. See Figure 8-3.

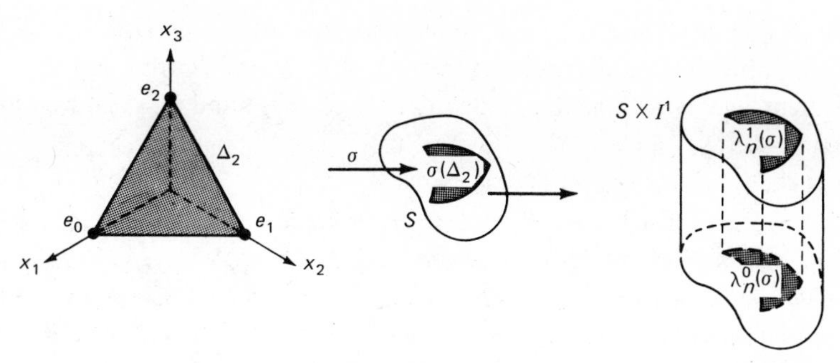

FIGURE 8-4

LEMMA. *If* $\alpha \in \mathscr{Z}_n(S, A)$, *then* $\lambda_n^1(\alpha) \sim \lambda_n^0(\alpha)$ *modulo* $A \times I^1$.

In case $\alpha = \sigma$ is a singular 2-simplex on S, our lemma requires that the singular simplexes whose images are contained in the lateral surface of the cylinder $S \times I^1$ in Figure 8-4 must cancel each other out.

THEOREM 8-6. *If* $A \subset S$, *then* $\lambda_*^0 = \lambda_*^1 : H_n(S, A) \to H_n(S \times I^1, A \times I^1)$.

Proof. If $[\alpha] \in H_n(S, A)$, then $\lambda_*^0([\alpha]) = [\lambda_n^0(\alpha)] = [\lambda_n^1(\alpha)] = \lambda_*^1([\alpha])$ from Definition 8-12 and the lemma above.

THEOREM 8-7 (Homotopy). *If* $f \simeq g : \langle S, A\rangle \to \langle T, B\rangle$, *then* $f_* = g_* : H_n(S, A) \to H_n(T, B)$.

Proof. Let h be a homotopy between f and g. This implies that $f = h \circ \lambda^0$ and $g = h \circ \lambda^1$. Hence $f_* = (h \circ \lambda^0)_* = h_* \circ \lambda_*^0 = h_* \circ \lambda_*^1 = (h \circ \lambda^1)_* = g_*$ by Theorem 8-6.

Our next results are consequences of the Homotopy Theorem. In particular, we show that homotopically equivalent topological pairs have isomorphic singular homology groups. This result implies that any deformation retract of a space S has singular homology groups isomorphic to those of S.

THEOREM 8-8. *If* $\langle S, A\rangle$ *and* $\langle T, B\rangle$ *are homotopically equivalent under the homotopy* f, *then* $f_* : H_n(S, A) \to H_n(T, B)$ *is an isomorphism onto.*

Proof. Let $g = f^{-1}$ and i and i' be identity mappings on $\langle S, A\rangle$ and $\langle T, B\rangle$, respectively. Thus $g \circ f \simeq i$ and $f \circ g \simeq i'$, which implies that $g_* \circ f_* = (g \circ f)_* = i_*$ and $f_* \circ g_* = (f \circ g)_* = i'_*$ by Theorem 8-7. Hence f_* and g_* are isomorphisms onto.

THEOREM 8-9. *If D is a deformation retract of S, then $H_n(D) \cong H_n(S)$.*

Proof. Since D is a deformation retract of S, D and S are homotopically equivalent. Applying Theorem 8-8 with $A = B = \varnothing$ and $T = D$, we obtain $H_n(D) \cong H_n(S)$.

THEOREM 8-10. *Let $B \subset A \subset S$ be spaces, and let $g : \langle S, B\rangle \to \langle S, A\rangle$ be the inclusion pair mapping. Let $f : \langle S, A\rangle \to \langle S, A\rangle$ be a pair mapping such that $f(A) \subset B$, and let $h : S \times I^1 \to S$ be a homotopy between f and the identity mapping on $\langle S, A\rangle$ such that $h(B \times I^1) \subset B$. Then $g_* : H_n(S, B) \to H_n(S, A)$ is an isomorphism onto.*

Proof. The mapping $h \mid \langle S \times I^1, B \times I^1\rangle$ is a homotopy between $f \mid \langle S, B\rangle$ and the identity mapping on $\langle S, B\rangle$. Moreover, $f \circ g = f \mid \langle S, B\rangle$ and $g \circ f = f$, which implies that $f : \langle S, A\rangle \to \langle S, B\rangle$ and $g : \langle S, B\rangle \to \langle S, A\rangle$ are homotopy inverses. Thus, from Theorem 8-8, we have $g_* : H_n(S, B) \to H_n(S, A)$ is an isomorphism onto.

Finally, we state the very important "Excision Theorem," which we shall find to be a most useful tool in the calculation of singular homology groups in the next two sections. We omit the proof, because it would require the introduction of several new ideas and results that are otherwise entirely peripheral to our discussion. By carefully studying Chapter 7 of Wallace's *An Introduction to Algebraic Topology*, the interested reader will find a detailed proof.

THEOREM 8-11 (Excision). *If $\bar{G} \subset U \subset A \subset S$, where U is open and $f : \langle S - G, A - G\rangle \to \langle S, A\rangle$ is the inclusion pair mapping, then $f_* : H_n(S - G, A - G) \to H_n(S, A)$ is an isomorphism onto for each integer n.*

EXERCISES

8-16. Determine $H_m(S)$ for all integers $m \geq 0$ in each case:

(a) $S = B^n$ (the solid n-dimensional unit ball).

(b) $S = I^n$ (the solid n-dimensional unit cube).

(c) $S = E^n$ (Euclidean n-space).

8-17. Show that the solid torus (doughnut), an annulus, and a circle all have the same singular homology groups.

8-18. Let $B = \{x \in E^n \mid |x| \leq 2\}$, $C = \{x \in E^n \mid 1 \leq |x| \leq 2\}$, and $S = \{x \in E^n \mid |x| = 2\}$. Thus C is a "collar" of the boundary S of the n-dimensional ball B. Show that $H_m(B, C) \cong H_m(B, S)$ for each integer m.

8-19. If

$$S^n = \left\{\langle x_1, \ldots, x_i, \ldots, x_{n+1}\rangle \in E^{n+1} \;\middle|\; \sum_{i=1}^{n+1} x_i^2 = 1\right\}$$

is the unit sphere in E^{n+1}, $A \subset S^n$ is the hemisphere for which $x_{n+1} \geq 0$, $A' \subset S^n$ is the subset for which $x_{n+1} \leq \frac{1}{2}$, $S^{n-1} \subset S^n$ is the subset for which $x_{n+1} = 0$, and $B = A \cap A'$, show that $H_m(S^n, A') \cong H_m(A, B) \cong H_m(A, S^{n-1})$ for each integer m.

8-20. Let S be the rectangular strip in Figure 8-5 with boundary A, and let T be the torus obtained by identifying the opposite edges. If $B = \alpha \cup \beta$, show that $H_n(S, A) \cong H_n(T, B)$.

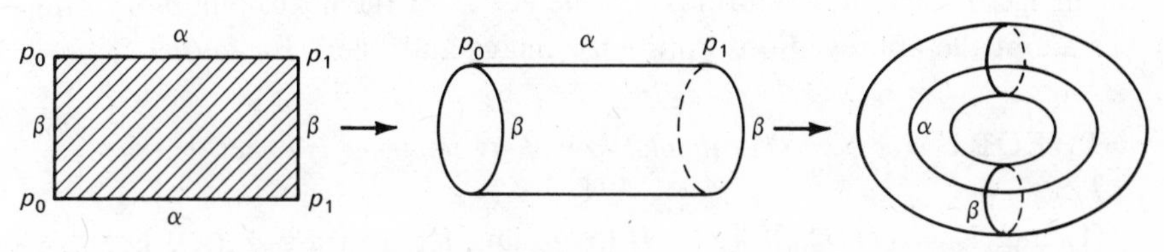

FIGURE 8-5

8-5 The Homology Sequence

In this section we define the singular homology sequence for a topological pair $\langle S, A\rangle$ and prove that it is "exact." The "exactness" of the homology sequence makes it one of our most useful tools in the calculation of singular homology groups. We preface this discussion with the definitions of the injection, projection, and boundary homomorphisms.

DEFINITION 8-14

(1) The *injection homomorphism* $i_* : H_n(A) \to H_n(S)$ is induced by the inclusion mapping $i : A \to S$. Thus $i_*([\alpha]) = [i_n(\alpha)] \in H_n(S)$ for each $[\alpha] \in H_n(A)$.

(2) The *projection homomorphism* $j_* : H_n(S) \to H_n(S, A)$ is induced by the inclusion pair mapping $j : \langle S, \varnothing\rangle \to \langle S, A\rangle$. Thus $j_*([\alpha]) = [j_n(\alpha)] \in H_n(S, A)$ for each $[\alpha] \in H_n(S)$.

(3) The *boundary homomorphism* $\delta : H_n(S, A) \to H_{n-1}(A)$ is defined by the rule $\delta([\alpha]) = [d\alpha] \in H_{n-1}(A)$ for each $[\alpha] \in H_n(S, A)$. (Show that δ is well defined.)

REMARK. To avoid confusion, we use the notation $i_*^m : H_m(A) \to H_m(S)$ and $i_*^n : H_n(A) \to H_n(S)$ when both injection homomorphisms are being considered in the same discussion. Similarly, we use the superscript notation j_*^m and j_*^n when necessary to avoid confusing two projection homomorphisms, and the superscript notation δ^m and δ^n when necessary to avoid confusing two boundary homomorphisms.

DEFINITION 8-15. The *singular homology sequence* of the topological pair $\langle S, A\rangle$ is the following sequence of groups and homomorphisms:

$$\cdots \xrightarrow{\delta^{n+1}} H_n(A) \xrightarrow{i_*{}^n} H_n(S) \xrightarrow{j_*{}^n} H_n(S, A) \xrightarrow{\delta^n} H_{n-1}(A) \xrightarrow{i_*{}^{n-1}} H_{n-1}(S) \xrightarrow{j_*{}^{n-1}} \cdots$$
$$\xrightarrow{\delta^1} H_0(A) \xrightarrow{i_*{}^0} H_0(S) \xrightarrow{j_*{}^0} H_0(S, A) \xrightarrow{\delta^0} \{0\}.$$

DEFINITION 8-16. The sequence defined in Definition 8-15 is *exact* iff the image of each homomorphism is the kernel of the next homomorphism. We adopt the abbreviations "img" for image and "ker" for kernel.

THEOREM 8-12. *The singular homology sequence is exact.*

Proof

(1) img i_*^n = ker j_*^n. If $[\alpha] \in H_n(S) \cap$ img i_*^n, then $\alpha \in \mathscr{Z}_n(A)$. Let $\beta = -\alpha \in \mathscr{C}_n(A)$ and $\gamma = 0$. Then $\alpha + \beta = d\gamma = 0$ and $\alpha \sim 0$ modulo A; i.e., $j_*^n([\alpha]) = 0$. This implies that img $i_*^n \subset$ ker j_*^n. On the other hand, $[\alpha] \in H_n(S)$ and $j_*^n([\alpha]) = 0$ imply that $\alpha = d\gamma + \beta$ for some $\beta \in \mathscr{C}_n(A)$ and some $\gamma \in \mathscr{C}_{n+1}(S)$. Thus $d\beta = d\alpha = 0$, which implies that $\beta \in \mathscr{Z}_n(A)$ and $[\beta] = [\alpha] \in$ img i_*^n. This shows that ker $j_*^n \subset$ img i_*^n, and thus img i_*^n = ker j_*^n.

(2) img j_*^n = ker δ^n. If $[\alpha] \in H_n(S, A) \cap$ img j_*^n, then $\alpha \in \mathscr{Z}_n(S)$ and $\delta^n([\alpha]) = 0$. Thus img $j_*^n \subset$ ker δ^n. On the other hand, if $[\alpha] \in H_n(S, A)$ and $\delta^n([\alpha]) = 0$, then $[d\alpha] = 0 \in H_{n-1}(A)$, which implies that $d\alpha \sim 0$ modulo A. Thus there exists $\beta \in \mathscr{C}_n(A)$ such that $d\alpha = d\beta$, which implies that $\alpha - \beta \in \mathscr{Z}_n(S)$ and $\alpha - \beta \sim \alpha$ modulo A. Hence $j_*^n([\alpha - \beta]) = [\alpha]$ and ker $\delta^n \subset$ img j_*^n. This shows that img j_*^n = ker δ^n.

(3) img δ^n = ker i_*^{n-1}. If $[\alpha] \in H_{n-1}(A) \cap$ img δ^n, then $\alpha = d\beta$ for some $\beta \in \mathscr{Z}_n(S, A)$. This implies that $\alpha \sim 0$ and $i_*^{n-1}([\alpha]) = 0$. Thus $[\alpha] \in$ ker i_*^{n-1} and img $\delta^n \subset$ ker i_*^{n-1}. On the other hand, if $[\alpha] \in H_{n-1}(A)$ and $i_*^{n-1}([\alpha]) = 0$, then $\alpha \sim 0$. This implies that $\alpha = d\beta$ for some $\beta \in \mathscr{C}_n(S)$. Since $\alpha \in \mathscr{C}_{n-1}(A)$, we have $\beta \in \mathscr{Z}_n(S, A)$ and $\delta^n([\beta]) = [d\beta] = [\alpha]$, which implies that $[\alpha] \in$ img δ^n. Thus ker $i_*^{n-1} \subset$ img δ^n, so that img δ^n = ker i_*^{n-1}.

THEOREM 8-13. *If* $f: \langle S, A\rangle \to \langle T, B\rangle$ *is a pair mapping, then* $\delta \circ f_* = (f \mid A)_* \circ \delta$.

Proof. If $[\alpha] \in H_n(S, A)$, then $\delta \circ f_*([\alpha]) = \delta[f_n(\alpha)] = [df_n(\alpha)] = [f_{n-1}(d\alpha)] = (f \mid A)_*([d\alpha]) = (f \mid A)_* \circ \delta([\alpha])$.

We conclude this section with several examples that demonstrate the usefulness of the Excision Theorem and the "exactness" of the homology sequence in the calculation of singular homology groups.

EXAMPLE 8-3. Let S denote the unit circle with points p_1, p_2, q_1, q_2, r_1, and r_2 chosen as shown in Figure 8-6. If A denotes the arc $p_1q_1r_1$ and B denotes the arc $p_2q_2r_2$, then $A \cup B = S$. Let $C = A \cap B$, which is the union of arc p_2p_1 and arc r_2r_1. By the Excision Theorem, $H_n(S, A) \cong H_n(B, C)$ for each integer n. Moreover, there exists a homotopy between the identity pair mapping on $\langle B, C\rangle$ and the pair mapping of $\langle B, C\rangle$ into itself which maps the arc p_2p_1 onto $\{p_2\}$ and maps the arc r_2r_1 onto $\{r_2\}$. By the Homotopy Theorem (8-7), we have $H_n(S, A) \cong H_n(B, C) \cong H_n(B, \{p_2, r_2\})$. By Exercise 8-23, $H_n(B, \{p_2, r_2\}) = \{0\}$ if $n \neq 1$ and $H_1(B, \{p_2r_2\}) \cong Z$.

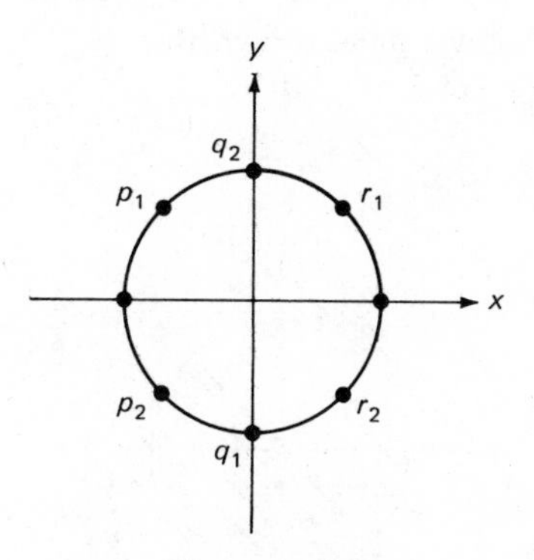

FIGURE 8-6

Consider now the homology sequence for the pair $\langle S, A\rangle$:

$$\longrightarrow H_n(A) \xrightarrow{i_*^n} H_n(S) \xrightarrow{j_*^n} H_n(S, A) \xrightarrow{\delta^n} H_{n-1}(A) \longrightarrow \cdots$$

$$\longrightarrow H_1(A) \xrightarrow{i_*^1} H_1(S) \xrightarrow{j_*^1} H_1(S, A) \xrightarrow{\delta^1} H_0(A) \xrightarrow{i_*^0} H_0(S) \xrightarrow{j_*^0} H_0(S, A).$$

If $n > 1$, $H_n(A) = \{0\}$, which implies that $\operatorname{img} i_*^n = \{0\} = \ker j_*^n$. Thus $H_n(S)$ is isomorphic to a subgroup of $H_n(S, A)$. From our remarks above, $H_n(S, A) = \{0\}$ and so $H_n(S) = \{0\}$ for $n > 1$. Moreover, $H_1(A) = \{0\}$, which implies that $\operatorname{img} i_*^1 = \{0\} = \ker j_*^1$. Thus $H_1(S)$ is isomorphic to a subgroup of the infinite cyclic group $H_1(S, A)$. Since $H_0(S, A) = \{0\}$, we have $\operatorname{img} i_*^0 = \ker j_*^0 = H_0(S)$. Since A and S are arcwise connected, $H_0(A) \cong H_0(S) \cong Z$ and i_*^0 is an isomorphism onto. This implies that $\operatorname{img} \delta^1 = \ker i_*^0 = \{0\}$, which implies that $\operatorname{img} j_*^1 = \ker \delta^1 = H_1(S, A)$. Hence j_*^1 is onto and $H_1(S)$ is infinite cyclic. Thus we have shown that $H_n(S) = \{0\}$ for $n > 1$ and $H_0(S) \cong H_1(S) \cong Z$.

EXAMPLE 8-4. Let S denote the closed unit circular disk with boundary A as shown in Figure 8-7. We consider the homology sequence for the pair $\langle S, A \rangle$:

$$H_n(A) \xrightarrow{i_*^n} H_n(S) \xrightarrow{j_*^n} H_n(S, A) \xrightarrow{\delta^n} H_{n-1}(A) \cdots H_2(A) \xrightarrow{i_*^2} H_2(S)$$
$$\xrightarrow{j_*^2} H_2(S, A) \xrightarrow{\delta^2} H_1(A) \xrightarrow{i_*^1} H_1(S) \xrightarrow{j_*^1} H_1(S, A) \xrightarrow{\delta^1} H_0(A).$$

For $n > 2$, $H_n(S) = \{0\}$, and also $H_{n-1}(A) = \{0\}$ from Example 8-3. Thus $\{0\} = \operatorname{img} j_*^n = \ker \delta^n = H_n(S, A)$ for $n > 2$. For $n = 2$, we have $H_2(S) = H_1(S) = \{0\}$, which implies that $\operatorname{img} \delta^2 = \ker i_*^1 = H_1(A)$ and $\{0\} = \operatorname{img} j_*^2 = \ker \delta^2$. Thus $H_2(S, A) \cong H_1(A)$, which is infinite cyclic from Example 8-3. Now if $\alpha \in \mathscr{Z}_1(S, A)$, then $d\alpha \sim 0$ modulo A, since $H_0(A)$ is infinite cyclic. This implies that $\operatorname{img} \delta^1 = \{0\}$, which implies that $H_1(S, A) = \ker \delta^1 = \operatorname{img} j_*^1 = \{0\}$, since $H_1(S) = \{0\}$. Moreover, $H_0(S, A) = \{0\}$ from Exercise 8-12. Thus we have shown that $H_n(S, A)$ is trivial for $n \neq 2$ and $H_2(S, A)$ is infinite cyclic.

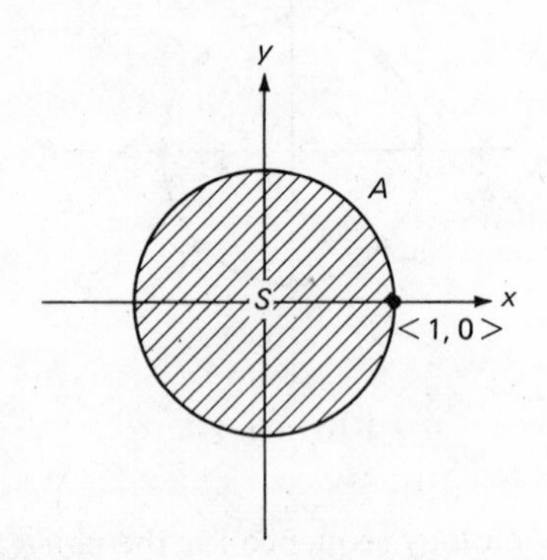

FIGURE 8-7

EXAMPLE 8-5. Let T be the torus described in Exercise 8-20 and pictured in Figure 8-5. Let $B = \alpha \cup \beta$. Then $H_n(B)$ is trivial for $n \geq 2$, $H_0(B)$ is infinite cyclic, and $H_1(B)$ is free Abelian with two generators by Exercise 8-24. Moreover, by the Excision Theorem $H_n(T, B) \cong H_n(S, A)$, where S is the closed unit circular disk with boundary A. From Example 8-4, we have $H_n(T, B)$ is trivial for $n \neq 2$ and $H_2(T, B)$ is infinite cyclic. Using this information and the "exactness" of the homology sequence of pair $\langle T, B \rangle$, one can show that $H_n(T)$ is trivial for $n \geq 3$, $H_0(T)$ and $H_2(T)$ are infinite cyclic and that $H_1(T)$ is free Abelian with two generators. We omit the details.

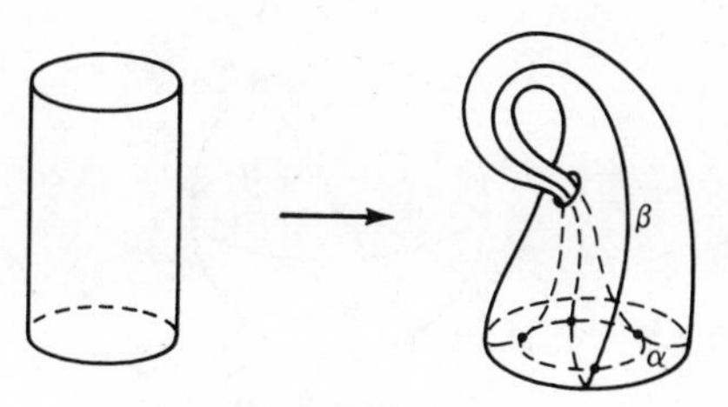

FIGURE 8-8

EXAMPLE 8-6. If we take a finite circular cylinder and identity the opposite ends while reversing the orientation of the two circles, the result is the "Klein bottle" B, a one-sided nonorientable surface that cannot be constructed in E^3. However, if we allow self-intersection, we can represent the Klein bottle in E^3 as in Figure 8-8. Although we omit the details, one can show that $H_n(B)$ is trivial for $n \geq 2$. From its construction and Figure 8-8, it is intuitively clear that $H_1(B)$ has two generators, a latitudinal free cycle α and a "twisted" longitudinal cycle β, which is a boundary 1-cycle if traversed an even number of times and a nonboundary 1-cycle if traversed an odd number of times. This intuitive argument suggests that $H_1(B) \cong Z \oplus Z/2$, where $Z/2$ is the group of integers modulo 2. Since B is arcwise connected, $H_0(B)$ is infinite cyclic.

EXAMPLE 8-7. The "Möbius strip" M is obtained by giving a rectangular strip a half-twist and then identifying the opposite ends. The result is the one-sided nonorientable surface shown in Figure 8-9. Although we omit the details, one can show that $H_n(M)$ is trivial if $n \geq 2$. Intuitively, if we traverse the path down the middle of M of a generating 1-cycle on M, we traverse a path that is homeomorphic to S^1. Thus $H_1(M)$ is infinite cyclic. Since M is arcwise connected, $H_0(M)$ is also infinite cyclic.

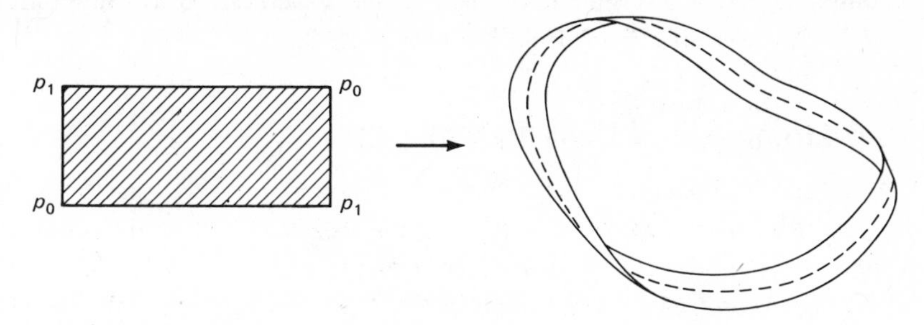

FIGURE 8-9

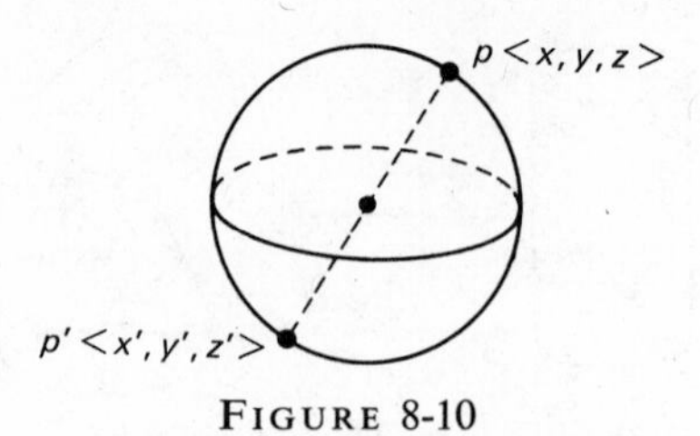

FIGURE 8-10

EXAMPLE 8-8. The "real projective plane" P^2 is obtained by identifying pairs of diametrically opposite points of $S^2 = \{\langle x, y, z\rangle \in E^3 \mid x^2 + y^2 + z^2 = 1\}$, as in Figure 8-10. Equivalently, $P^2 = \{\langle x, y, z\rangle \in E^3 \mid \langle x, y, z\rangle \neq \langle 0, 0, 0\rangle$ and $\langle x, y, z\rangle = \langle x', y', z'\rangle$ iff $\exists \lambda \neq 0$ such that $x = \lambda x'$, $y = \lambda y'$, $z = \lambda z'\}$. One can show that $H_n(P^2)$ is trivial for $n \geq 2$, $H_0(P^2)$ is infinite cyclic and $H_1(P^2) \cong Z/2$. Details can be found in Wallace, *An Introduction to Algebraic Topology*, p. 159.

EXERCISES

8-21. Verify that the mapping $\delta : H_n(S, A) \to H_{n-1}(A)$ given by $\delta([\alpha]) = [d\alpha]$ in Definition 8-14 actually is a homomorphism by showing that $\delta([\alpha] + [\beta]) = \delta([\alpha]) + \delta([\beta])$ for all $[\alpha], [\beta] \in H_n(S, A)$.

8-22. If $A = \{p\}$ with $p \in S$, use the "exactness" of the homology sequence to show that $H_n(S, A) \cong H_n(S)$ for each integer $n \geq 1$.

8-23. Let $S = [0, 1]$ and $A = \{0, 1\}$. Use the "exactness" of the homology sequence to prove that $H_n(S, A) = \{0\}$ if $n \neq 1$ and $H_1(S, A) \cong Z$.

8-24. If S is a subspace of E^2 composed of two circles with exactly one point in common, use the Excision Theorem and the "exactness" of the homology sequence to determine $H_n(S)$ for all integers $n \geq 0$.

8-6 Homology of Cells and Spheres

As a further application of the Excision Theorem and the "exactness" of the homology sequence, we calculate for all $m \geq 0$ the singular homology groups of the n-sphere S^n and also those of the pair $\langle B^n, S^{n-1}\rangle$, where B^n is the (closed) unit n-cell with boundary S^{n-1}. Our calculation embodies the technique used in Example 8-3 and is essentially that of Wallace (see pp. 162–66).

EXAMPLE 8-9. Consider the following portion of the homology sequence for the pair $\langle B^n, S^{n-1}\rangle$ where $n, m \geq 2$:

$$\cdots \xrightarrow{\delta^{m+1}} H_m(S^{n-1}) \xrightarrow{i_*{}^m} H_m(B^n) \xrightarrow{j_*{}^m} H_m(B^n, S^{n-1})$$
$$\xrightarrow{\delta^m} H_{m-1}(S^{n-1}) \xrightarrow{i_*{}^{m-1}} H_{m-1}(B^n) \xrightarrow{j_*{}^{m-1}} \cdots.$$

Recall that $H_m(B^n)$ and $H_{m-1}(B^n)$ are trvial for $m \geq 2$ (Exercise 8-16). Hence $\ker \delta^m = \text{img}\, j_*^m = \{0\}$ and $\text{img}\, \delta^m = \ker i_*^{m-1} = H_{m-1}(S^{n-1})$. Thus δ^m is an isomorphism onto; i.e., for $n, m \geq 2$ we have

$$H_m(B^n, S^{n-1}) \cong H_{m-1}(S^{n-1}). \tag{1}$$

For $n \geq 2$ and $m = 1$, we consider the following portion of the homology sequence of $\langle B^n, S^{n-1}\rangle$:

$$\cdots \xrightarrow{\delta^2} H_1(S^{n-1}) \xrightarrow{i_*{}^1} H_1(B^n) \xrightarrow{j_*{}^1} H_1(B^n, S^{n-1}) \xrightarrow{\delta^1} H_0(S^{n-1})$$
$$\xrightarrow{i_*{}^0} H_0(B^n) \xrightarrow{j_*{}^0} H_0(B^n, S^{n-1}) \xrightarrow{\delta^0} \{0\}.$$

Since S^{n-1} and B^n are arcwise connected, $H_0(S^{n-1})$ and $H_0(B^n)$ are infinite cyclic ($n \geq 2$). Thus i_*^0 is an isomorphism onto and $\text{img}\, \delta^1 = \ker i_*^0 = \{0\}$. This implies that j_*^1 is onto, since $\text{img}\, j_*^1 = \ker \delta^1 = H_1(B^n, S^{n-1})$. Since $H_1(B^n) = \{0\}$, we have for $n \geq 2$:

$$H_1(B^n, S^{n-1}) = \{0\}. \tag{2}$$

Since B^n is arcwise connected, for $n \geq 1$ it follows from Exercise 8-12 that

$$H_0(B^n, S^{n-1}) = \{0\}. \tag{3}$$

Next, we employ the technique used in Example 8-3. Observe that S^n is contained in $A_1^n \cup A_2^n$, where $A_1^n = \{\langle x_1, \ldots, x_{n+1}\rangle \in S^n \,|\, x_{n+1} \geq -\varepsilon\}$ and $A_2^n = \{\langle x_1, \ldots, x_{n+1}\rangle \,|\, x_{n+1} \leq \varepsilon\}$. If we let $A = A_1^n \cap A_2^n$, then for $n \geq 1$ and all $m \geq 0$ we have as a consequence of the Excision Theorem:

$$H_m(S^n, A_2^n) \cong H_m(A_1^n, A). \tag{4}$$

If we let $C = \{\langle x_1, \ldots, x_{n+1}\rangle \in S^n \mid x_{n+1} = -\varepsilon\}$, and let f be a pair mapping of $\langle A_1^n, A\rangle$ into itself such that f is homotopic to the identity pair mapping of $\langle A_1^n, A\rangle$ and $f(A) = C$, then $H_m(A_1^n, A) \cong H_m(A_1^n, C)$ by Theorem 8-8. Since A_1^n is homeomorphic to B^n and C is homeomorphic to S^{n-1}, it follows that

$$H_m(A_1^n, A) \cong H_m(B^n, S^{n-1}). \tag{5}$$

As a consequence of (4) and (5), we have for $n \geq 1$ and for all $m \geq 0$:

$$H_m(S^n, A_2^n) \cong H_m(B^n, S^{n-1}). \tag{6}$$

Now consider the following portion of the homology sequence of $\langle S^n, A_2^n \rangle$ for $n \geq 1$:

$$\cdots \xrightarrow{\delta^{m+1}} H_m(A_2^n) \xrightarrow{i_*^m} H_m(S^n) \xrightarrow{j_*^m} H_m(S^n, A_2^n)$$
$$\xrightarrow{\delta^m} H_{m-1}(A_2^n) \xrightarrow{i_*^{m-1}} H_{m-1}(S^n) \xrightarrow{j_*^{m-1}} \cdots.$$

From Exercise 8-16 we have that $H_m(A_2^n)$ and $H_{m-1}(A_2^n)$ are trivial for $m \geq 2$, since A_2^n is a (closed) n-cell. This implies that j_*^m is an isomorphism onto, since $\ker j_*^m = \operatorname{img} i_*^m = \{0\}$ and $\operatorname{img} j_*^m = \ker \delta^m = H_m(S^n, A_2^n)$. Moreover, if $m = 1$, then i_*^0 is an isomorphism onto, since $H_0(A_2^n)$ and $H_0(S^n)$ are both infinite cyclic (A_2^n and S^n are arcwise connected). This implies that $\ker i_*^0 = \operatorname{img} \delta^1 = \{0\}$, which implies that $\operatorname{img} j_*^1 = \ker \delta^1 = H_1(S^n, A_2^n)$. Also, $\ker j_*^1 = \operatorname{img} i_*^1 = \{0\}$, since $H_1(A_2^n) = \{0\}$. Thus j_*^1 is an isomorphism onto, and we obtain for $n, m \geq 1$:

$$H_m(S^n) \cong H_m(S^n, A_2^n). \tag{7}$$

Combining (6) and (7), we obtain for $n, m \geq 1$:

$$H_m(S^n) \cong H_m(B^n, S^{n-1}). \tag{8}$$

Combining (1) and (8), we obtain for $n, m \geq 2$:

$$H_m(B^n, S^{n-1}) \cong H_{m-1}(B^{n-1}, S^{n-2}). \tag{9}$$

For $m > n$, $n - 1$ successive applications of (9) yield $H_m(B^n, S^{n-1}) \cong H_{m-n+1}(B^1, S^0) = \{0\}$ by Exercise 8-23. For $m < n$, $m - 1$ successive applications of (9) yield $H_m(B^n, S^{n-1}) \cong H_1(B^{n-m+1}, S^{n-m}) = \{0\}$ by (2). For $m = n$, $m - 1$ successive applications of (9) yield $H_m(B^n, S^{n-1}) \cong H_1(B^1, S^0) \cong Z$ from Exercise 8-23.

In summary, $H_m(B^n, S^{n-1}) = \{0\}$ for $m \neq n$ and $H_n(B^n, S^{n-1}) \cong Z$ for $n, m \geq 2$. From (3), $H_0(B^n, S^{n-1}) = \{0\}$ for $n \geq 1$. From (2), $H_1(B^n, S^{n-1}) = \{0\}$ for $n \geq 2$. Clearly, $H_n(B^1, S^0) = \{0\}$ for $m \geq 1$ and $H_1(B^1, S^0) \cong Z$. Using these results in conjunction with (8), we obtain $H_m(S^n) = \{0\}$ for $m \neq n$ and $H_n(S^n) \cong Z$ for $n, m \geq 1$. Since S^n is arcwise connected for $n \geq 1$, we have $H_0(S^n) \cong Z$. Clearly, $H_m(S^0) = \{0\}$ for $m \geq 1$ and $H_0(S^0) \cong Z \oplus Z$.

EXERCISE

8-25. Supply the details necessary to demonstrate that $H_m(S^0) = \{0\}$ for $m \geq 1$ and $H_0(S^0) \cong Z \oplus Z$.

8-7 Applications of Homotopy and Homology

We conclude our discussion of algebraic topology by discussing some elegant applications, which hopefully will reward the reader for his efforts up to this point. We introduce the notion of the "homological degree" of a mapping f of one n-sphere into another. Using this concept, together with results from homotopy theory and singular homology theory, we are able to construct simple proofs of the Fundamental Theorem of Algebra, the No-Retraction Theorem, and the Brouwer fixed-point theorem, whose classical proofs are somewhat long and tedious.

DEFINITION 8-17. Let $n \geq 1$ and let $f: S^n \to S^n$ be a mapping. Then f induces a homomorphism $f_* : H_n(S^n) \to H_n(S^n)$, where $H_n(S^n)$ is infinite cyclic from Example 8-9. Let $[\gamma]$ be a generator of $H_n(S^n)$. Then $f_*([\gamma]) \in H_n(S^n)$, and there exists an integer k such that $f_*([\gamma]) = k[\gamma]$. Moreover, if $[\alpha] \in H_n(S^n)$ is arbitrary, there exists an integer m such that $[\alpha] = m[\gamma]$. Hence $f_*([\alpha]) = f_*(m[\gamma]) = mf_*([\gamma]) = mk[\gamma] = km[\gamma] = k[\alpha]$. This uniquely determined integer k is called the *homological degree* of f. Thus f is of degree k iff $f_*(x) = kx$ for each $x \in H_n(S^n)$. We use the abbreviation "deg (f)" for the homological degree of f.

Although the notion of "homological degree" is different from that of "polynomial degree," we demonstrate their compatibility with the following example and theorem.

EXAMPLE 8-10. Let $n = 1$ and consider $f: S^1 \to S^1$ given by $f(z) = a$ constant c, where z is a complex number $\langle x, y\rangle$ with modulus $|z| = \sqrt{x^2 + y^2} = 1$. Then $f_* : H_1(S^1) \to H_1(S^1)$, where $H_1(S^1) \cong Z$ is the induced homomorphism. If $[\gamma]$ is a generator of $H_1(S^1)$, then $\gamma = \sum_{i=1}^{l} m_i\sigma_i$, where $m_i \in Z$, $\sigma_i \in \mathscr{S}_n(S^1)$ $(i = 1, \ldots, l)$. Hence

$$
\begin{aligned}
f_*([\gamma]) &= [f_1(\gamma)] \\
&= \left[\sum_{i=1}^{l} m_i f_1(\sigma_i)\right] = \left[c \sum_{i=1}^{l} m_i\right] = [0] = k[\gamma] \qquad \text{iff } k = 0.
\end{aligned}
$$

Thus the homological degree of f is 0, which is also the polynomial degree of f. Next, consider $g : S^1 \to S^1$ given by $g(z) = z$ for each complex number $z = \langle x, y\rangle$ with modulus $|z| = \sqrt{x^2 + y^2} = 1$. Then

$$
\begin{aligned}
g_*([\gamma]) &= [g_1(\gamma)] \\
&= \left[\sum_{i=1}^{l} m_i g_1(\sigma_i)\right] = \left[\sum_{i=1}^{l} m_i\sigma_i\right] = [\gamma] = k[\gamma] \qquad \text{iff } k = 1.
\end{aligned}
$$

Thus the homological degree of g is 1, as is its polynomial degree.

THEOREM 8-14. *If $f_n : S^1 \to S^1$ is the mapping given by $f_n(z) = z^n$ for each complex number $z \in S^1$, then the induced homomorphism $f_{n*} : H_1(S^1) \to H_1(S^1)$ is given by $f_{n*}(x) = nx$ for each $x \in H_1(S^1)$, and thus f_n has homological degree n.*

Proof. The result has already been established for $n = 0$ and $n = 1$ in Example 8-10. The proof for the general case is by induction and is essentially that found in Hu's *Homology Theory*, pp. 97–98. Assume that the result is true for some integer $k \geq 1$ and let $f_{k+1} : S^1 \to S^1$ be given by $f_{k+1}(z) = z^{k+1}$ for each complex number $z \in S^1$. We define mappings g and h from S^1 into S^1 as follows:

$$g(z) = g(e^{2\pi i\theta}) = \begin{cases} e^{2(k+1)\pi i\theta} & \text{if } 0 \leq \theta \leq \dfrac{k}{k+1} \\ 1 & \text{if } \dfrac{k}{k+1} \leq \theta \leq 1 \end{cases}$$

$$h(z) = h(e^{2\pi i\theta}) = \begin{cases} 1 & \text{if } 0 \leq \theta \leq \dfrac{k}{k+1} \\ e^{2(k+1)\pi i\theta} & \text{if } \dfrac{k}{k+1} \leq \theta \leq 1. \end{cases}$$

It can be shown that $f_{(k+1)*} = g_* + h_*$, where these mappings are the homomorphisms of $H_1(S^1)$ into $H_1(S^1)$ induced by f_{k+1}, g, and h. Moreover, the function $G : S^1 \times I^1 \to S^2$ given by

$$G(z, t) = G(e^{2\pi i\theta}, t) = \begin{cases} e^{2\pi(k+1-t)i\theta} & \text{if } 0 \leq \theta \leq \dfrac{k}{k+1}, \quad t \in I^1 \\ 1 & \text{if } \dfrac{k}{k+1} \leq \theta \leq 1, \quad t \in I^1 \end{cases}$$

is a homotopy between $G(z, 0) = g(z)$ and $G(z, 1) = z^k = f_k(z)$. Also, the function $H : S^1 \times I^1 \to S^1$ given by

$$H(z, t) = H(e^{2\pi i\theta}, t) = \begin{cases} 1 & \text{if } 0 \leq \theta \leq \dfrac{k}{k+1}, \quad t \in I^1 \\ e^{2\pi[t+(k+1)(1-t)]i\theta} & \text{if } \dfrac{k}{k+1} \leq \theta \leq 1, \quad t \in I^1 \end{cases}$$

is a homotopy between $H(z, 0) = h(z)$ and $H(z, 1) = z = f_1(z)$. Thus, by the Homotopy Theorem (8-7), we have $f_{(k+1)*} = g_* + h_* = f_{k*} + f_{1*}$. Moreover, if $x \in H_1(S^1)$, then $f_{(k+1)*}(x) = f_{k*}(x) + f_{1*}(x) = k \cdot x + 1 \cdot x = (k + 1) \cdot x$. Thus, by Definition 8-17, the homological degree of f_{k+1} is $k + 1$, as is its polynomial degree.

COROLLARY. *If* $f_n : S^n \to S^n$ *is the mapping given by* $f_n(z) = z^n$ *for each complex number* $z \in S^1$, *then the induced homomorphism* $f_{n*} : H_n(S^n) \to H_n(S^n)$ *is given by* $f_{n*}(x) = nx$ *for each* $x \in H_n(S^n)$, *and thus* f_n *has homological degree* n.

We now show that homotopic mappings have the same homological degree, from which we deduce two interesting corollaries concerning mappings of n-spheres. The converse (due to Hopf) is also true, although we do not prove it here. The interested reader will find a rather elegant proof of it (using the "Freudenthal suspension" technique of simplicial homology) in Dugundji's *Topology*, pp. 350–52.

THEOREM 8-15. *If* $f \simeq g : S^n \to S^n$, *then* $\deg(f) = \deg(g)$.

Proof. Let $f \simeq g$ and suppose that $\deg(f) = k$. We show that $\deg(g) = k$ also. If $f_* : H_n(S^n) \to H_n(S^n)$ and $g_* : H_n(S^n) \to H_n(S^n)$ are the induced homomorphisms, then $f_* = g_*$ by the Homotopy Theorem (8-7); i.e., $f_*(x) = g_*(x)$ for each $x \in H_n(S^n)$. Now let $[\gamma]$ be any generator of $H_n(S^n)$. Then $g_*([\gamma]) = f_*([\gamma]) = k[\gamma]$, since $\deg(f) = k$. Hence $\deg(g) = k$.

COROLLARY 1. S^n *is not contractible.*

Proof. Suppose S^n were contractible. Then the identity mapping on S^n would be homotopic to a constant mapping of S^n. However, this is impossible in view of Theorem 8-15, since the identity mapping has degree 1 and a constant mapping has degree 0 (Example 8-10).

COROLLARY 2. *If* $f : S^n \to S^n$ *is continuous and* $\deg(f) \neq 0$, *then* $f(S^n) = S^n$.

Proof. Suppose that $S^n - f(S^n) \neq \varnothing$ and let $p \in S^n - f(S^n)$. Then $S^n - \{p\}$ is contractible; i.e., the identity mapping on $S^n - \{p\}$ is homotopic to a constant mapping of $S^n - \{p\}$. Thus f is homotopic to a constant mapping, since $f(S^n) \subset S^n - \{p\}$. This implies that $\deg(f) = 0$, which is a contradiction.

Now we are ready to proceed with the proof of the Fundamental Theorem of Algebra. The polynomial $P(z) = a_0 + a_1 z + \cdots + a_{n-1}z^{n-1} + z^n$, where $n \in I^+$ and a_i is a complex number ($i = 0, 1, \ldots, n - 1$), is a mapping of E^2 into E^2. If we let $\bar{P}(\infty) = \infty$ and observe that $S^2 = E^2 \cup \{\infty\}$ (the "one-point compactification" of E^2), we may regard $\bar{P}$ as an extension of P to S^2. To see that $\bar{P}$ is continuous at ∞, let $G \subset S^2$ be open and $\infty \in G$. Then $S^2 - G$ is a compact subset of E^2 (thus closed and bounded). Hence there exists $A > 0$ such that $S^2 - G \subset \{z \mid |z| \leq A\}$. Since $P(z)$ is a polynomial, there exists $B > 0$ such that $|P(z)| > A$ for all $z \in E^2$ such that $|z| > B$. Thus the set $V = \{z \in S^2 \mid |z| > B\}$ is an open set containing ∞ such that $\bar{P}(V) \subset G$, as required for continuity at ∞. We shall need the following lemma in our proof of the Fundamental Theorem of Algebra.

LEMMA. *$\bar{P}(z)$ is homotopic to $\bar{f}_n(z)$, the continuous extension of $f_n(z) = z^n$ to S^2 obtained by letting $\bar{f}_n(\infty) = \infty$ and $\bar{f}_n(z) = f_n(z)$ if $z \in E^2$.*

Proof. Let $h(z, t) = z^n + (1 - t)(a_0 + a_1 z + \cdots + a_{n-1}z^{n-1})$ if $z \in E^2$ and $h(\infty, t) = \infty$ for all $t \in I^1$. Clearly, h is continuous for all $t \in I^1$ and $z \in E^2$. To see that h is continuous at $\langle \infty, t \rangle$ for all $t \in I^1$, let $A > 0$ be arbitrary and let $M = \sum_{i=1}^{n-1} |a_i|$. Then for all $t \in I^1$ and each $z \in E^2$ such that $|z| > B = \max\{A + M, 1\}$,

$$\begin{aligned} |h(z, t)| &\geq |z|^n - (1 - t)(|a_0| + |a_1|\,|z| + \cdots + |a_{n-1}|\,|z|^{n-1}) \\ &\geq |z|^n - M|z|^{n-1} \\ &\geq |z| - M \\ &> (A + M) - M = A. \end{aligned}$$

Thus $\lim_{z\to\infty} h(z, t) = \infty = h(\infty, t)$ for all $t \in I^1$, as required for continuity at $\langle \infty, t \rangle$. Moreover, $h(z, 0) = \bar{P}(z)$ and $h(z, 1) = z^n = \bar{f}_n(z)$ for all $z \in S^2$. Hence h is the required homotopy between $\bar{P}$ and $\bar{f}_n$.

THEOREM 8-16 (Fundamental Theorem of Algebra). *The polynomial $\bar{P}(z)$ has at least one complex zero.*

Proof. By our lemma and Theorem 8-15, $\bar{P}$ and $\bar{f}_n$ have the same homological degree. Thus, as a consequence of the corollary to Theorem 8-14, we have that $\deg(\bar{P}) = \deg(\bar{f}_n) = n$. Moreover, since $n > 0$, $\bar{P}(S^2) = S^2$ by Corollary 2 to Theorem 8-15. Hence there exists some $z \in E^2$ such that $\bar{P}(z) = P(z) = 0$.

Another classical application of homotopy and homology involves showing that E^m and E^n are nonhomeomorphic for $m \neq n$ $(m, n \in I^+)$. This is done by contradiction. One first shows that S^m and S^n are not homotopically equivalent for $m \neq n$. Next, one shows that $E^m - \{0\}$ and $E^n - \{0\}$ are homotopically equivalent to S^{m-1} and S^{n-1}, respectively. Thus, if E^m and E^n were homeomorphic, then $E^m - \{0\}$ and $E^n - \{0\}$ would be homotopically equivalent. This would imply that S^{m-1} and S^{n-1} were homotopically equivalent, which is impossible, since $m - 1 \neq n - 1$. Details can be found in Hu's *Holomogy Theory*, pp. 150–52.

We continue our discussion of applications of homotopy and homology by proving the "No-Retraction Theorem," which states that there is no retraction of an n-cell into its boundary if $n > 0$. This result is then used to prove the classical "Brouwer fixed-point theorem." In fact, the two results are equivalent.

THEOREM 8-17 (No-Retraction Theorem). *There is no retraction of an n-cell $B^n (n > 0)$ onto its boundary S^{n-1}.*

Proof. Suppose that r were a retraction of B^n onto S^{n-1}. Let $h(x, t) = r[(1 - t)x]$ for each $x \in S^{n-1}$. Thus $h(x, 0) = r(x)$ and $h(x, 1) = r(0) \in S^{n-1}$. Clearly, h is a homotopy between the identity mapping on S^{n-1} and the constant mapping of S^{n-1} onto $\{r(0)\}$. This implies that S^{n-1} is contractible, which contradicts Corollary 1 to Theorem 8-15.

DEFINITION 8-18. If $f: S \to S$ is a mapping from S into S, then $x \in S$ is a *fixed point* of f iff $f(x) = x$. Moreover, a space S has the *fixed-point property* iff each mapping $f: S \to S$ has a fixed point.

EXAMPLE 8-11. Consider the linear function f given by $f(x) = mx + b$ for each real number x with $m \neq 0$. If $m \neq 1$, then f has the unique fixed point $x_0 = b/(1 - m)$. If $m = 1$, then f has no fixed point if $b \neq 0$ and has each real number as a fixed point if $b = 0$.

EXAMPLE 8-12. I^1 has the fixed-point property. Let f be a mapping of I^1 into itself and suppose that $f(x) \neq x$ for each $x \in I^1$. Then the function g given by $g(x) = f(x) - x$ is continuous, and thus $g(I^1)$ must be connected, since I^1 is connected. Also, $g(0) = f(0) > 0$ and $g(1) = f(1) - 1 < 0$. Hence there exists $x_0 \in I^1$ such that $g(x_0) = f(x_0) - x_0 = 0$ by Theorem 4-6. This implies that $f(x_0) = x_0$ is a fixed point, which contradicts our assumption that f has no fixed points.

We leave to the reader the exercise of showing that the fixed-point property is a topological property; i.e., it is invariant under homeomorphisms. However, we do prove that the fixed-point property is invariant under retractions. Then we conclude with the statement and proof of the Brouwer fixed-point theorem.

THEOREM 8-18. *The fixed-point property is invariant under retractions.*

Proof. Suppose S has the fixed-point property, and let $r: S \to A$ be a retraction of S onto a subspace A of S. If $f: A \to A$ is any mapping of A

into itself, then the composition $i \circ f \circ r : S \to S$ is continuous, where $i : A \to S$ is the inclusion mapping. Thus there exists $x_0 \in S$ such that $i \circ f \circ r(x_0) = x_0$, since S has the fixed-point property. This implies that $i \circ f \circ r(x_0) = f[r(x_0)] = x_0 \in A$. Hence A has the fixed-point property.

THEOREM 8-19 (Brouwer Fixed-Point Theorem). *The closed n-cell B^n has the fixed-point property.*

Proof. Let $f: B^n \to B^n$ be a mapping of B^n into itself, and suppose that $f(x) \neq x$ for every $x \in B^n$. For each $x \in B^n$, let $L(x)$ denote the directed ray from $f(x)$ to x. Denote by $r(x)$ the point of intersection of $L(x)$ and the boundary S^{n-1} of B^n. This defines a function $r : B^n \to S^{n-1}$, such that $r(x) = x$ for each $x \in S^{n-1}$, which is continuous, since f is continuous. This r is a retraction of B^n onto its boundary S^{n-1}, which contradicts Theorem 8-17. Hence there exists some $x \in B^n$ such that $f(x) = x$, as required.

EXERCISES

8-26. If f and g are mappings of S^n into itself, show that $\deg(f \circ g) = \deg(f) \cdot \deg(g)$.

8-27. Explain how our proof of the Fundamental Theorem of Algebra breaks down if $P(z)$ is replaced by the polynomial $P(x)$ with $x \in E^1$.

8-28. Show that the fixed-point property is a topological property.

8-29. Prove that S^n does not have the fixed-point property for each integer $n \geq 0$.

8-30. Show that the n-dimensional unit cube I^n and the Hilbert cube I^ω each have the fixed-point property.

8-31. Prove that the Brouwer fixed-point theorem implies the No-Retraction Theorem and hence is equivalent to it.

Bibliography

Books

BING, R. H. *Elementary Point Set Topology* (Slaught Memorial Paper 8). Washington, D.C.: Mathematical Association of America, 1960.

BIRKHOFF, G. *Lattice Theory*, rev. ed. (Colloquium Publication 25). Providence, R.I.: American Mathematical Society, 1948.

BOURBAKI, N. *Elements of Mathematics* (Parts 1 and 2). Reading, Mass.: Addison-Wesley Publishing Co., Inc., 1966.

COHEN, P. J. *Set Theory and the Continuum Hypothesis.* Menlo Park, Calif.: W. A. Benjamin, Inc., 1966.

COPSON, E. T. *Metric Spaces* (Cambridge Tracts in Mathematics 57). New York: Cambridge University Press, 1968.

DUGUNDJI, J. *Topology.* Boston: Allyn and Bacon, Inc., 1968.

EILENBERG, S., and N. STEENROD. *Foundations of Algebraic Topology.* Princeton, N.J.: Princeton University Press, 1952.

GEMIGNANI, M. C. *Elementary Topology.* Reading, Mass.: Addison-Wesley Publishing Co., Inc., 1967.

GILLMAN, L., and M. JERISON. *Rings of Continuous Functions.* New York: Van Nostrand Reinhold Company, 1960.

GREEVER, J. *Theory and Examples of Point-Set Topology.* Belmont, Calif.: Brooks/Cole, Belmont, 1967.

HALL, D. W., and G. L. SPENCER. *Elementary Topology.* New York: John Wiley & Sons, Inc., 1955.

HALMOS, P. R. *Naïve Set Theory.* New York: Van Nostrand Reinhold Company, 1960.

HOCKING, J. G., and G. S. YOUNG. *Topology.* Reading, Mass.: Addison-Wesley Publishing Co., Inc., 1961.

HU, S. T. *Elements of General Topology.* San Francisco: Holden-Day, Inc., 1964.

———. *Homology Theory.* San Francisco: Holden-Day, Inc., 1966.

———. *Homotopy Theory.* New York: Academic Press, Inc., 1959.

———. *Theory of Retracts.* Detroit, Mich.: Wayne State University Press, 1965.

HUREWICZ, W., and H. WALLMAN. *Dimension Theory.* Princeton, N.J.: Princeton University Press, 1941.

KELLEY, J. L. *General Topology.* New York: Van Nostrand Reinhold Company, 1955.

KOLMOGOROV, A. N., and S. V. FOMIN. *Elements of the Theory of Functions and Functional Analysis*, Vol. 1: *Metric and Normed Spaces.* Rochester, N.Y.: Graylock Press, 1957.

MASSEY, W. S. *Algebraic Topology: An Introduction.* New York: Harcourt Brace Jovanovich, Inc., 1967.

MOORE, R. L. *Foundations of Point Set Theory*, rev. ed. (Colloquium Publication 13). Providence, R.I.: American Mathematical Society, 1962.

PERVIN, W. J. *Foundations of General Topology.* New York: Academic Press, Inc., 1964.

SCHUBERT, H. *Topology.* Boston: Allyn and Bacon, Inc., 1968.

SIERPINSKI, W. *General Topology.* Toronto: University of Toronto Press, 1952.

STEEN, L. A., and J. A. SEEBACH. *Counterexamples in Topology.* New York: Holt, Rinehart and Winston, Inc., 1970.

Summer Institute on Set Theoretic Topology, rev. ed. Madison, Wisc., 1957.

THRON, W. J. *Topological Structures.* New York: Holt, Rinehart and Winston, Inc., 1966.

TUKEY, J. W. *Convergence and Uniformity in Topology* (Annals of Mathematics Studies 2). Princeton, N.J.: Princeton University Press, 1940.

WALLACE, A. H. *An Introduction to Algebraic Topology.* Elmsford, N.Y.: Pergamon Press, Inc., 1957.

WHYBURN, G. T. *Topological Analysis.* Princeton, N.J.: Princeton University Press, 1958.

WILDER, R. L. *Topology of Manifolds* (Colloquium Publication 32). Providence R.I.: American Mathematical Society, 1949.

WILLARD, S. *General Topology.* Reading, Mass.: Addison-Wesley Publishing Co., Inc., 1970.

Papers

ALEXANDROFF, P. "Some Results in the Theory of Topological Spaces, Obtained Within the Last Twenty-five Years." *Russian Math. Surveys*, **15** (1960), 23–83.

ANDERSON, R. D. "Hilbert Space Is Homeomorphic to the Countable Infinite Product of Lines." *Bull. Amer. Math. Soc.*, **72** (1966), 515–19.

ARENS, R., and J. DUGUNDJI. "Topologies for Function Spaces." *Pac. J. Math.*, **1** (1951), 5–31.

AULL, C. E., and W. J. THRON. "Separation Axioms Between T_0 and T_1." *Indag. Math.*, **24** (1963), 26–37.

BARTLE, R. G. "Nets and Filters in Topology." *Amer. Math. Monthly*, **62** (1955), 551–57.

BELL, H. "On Fixed Point Properties of Plane Continua." *Trans. Amer. Math. Soc.*, **128** (1967), 539–48.

BING, R. H. "A Countable Connected Hausdorff Space." *Proc. Amer. Math. Soc.*, **4** (1953), 474.

———. "A Translation of the Normal Moore Space Conjecture." *Proc. Amer. Math. Soc.*, **16** (1965), 612–19.

———. "Extending a Metric." *Duke Math. J.*, **14** (1947), 511–19.

———. "Metrization of Topological Spaces." *Can. J. Math.*, **3** (1951), 175–86.

———. "The Elusive Fixed-Point Property." *Amer. Math. Monthly*, **76** (1969), 119–32.

ČECH, E. "On Bicompact Spaces." *Ann. Math.*, **38** (1937), 823–44.

CEDAR, J. "Some Generalizations of Metric Spaces." *Pac. J. Math.*, **11** (1961), 105–25.

CHITTENDEN, E. W. "On the Metrization Problem and Related Problems in the Theory of Abstract Sets." *Bull. Amer. Math. Soc.*, **33** (1927), 13–34.

DAVIS, A. S. "Indexed Systems of Neighborhoods for General Topological Spaces." *Amer. Math. Monthly*, **68** (1961), 886–93.

DUGUNDJI, J. "An Extension of Tietze's Theorem." *Pac. J. Math.*, **1** (1951), 353–67.

EILENBERG, S. "Singular Homology Theory." *Ann. Math.*, **45** (1944), 407–47.

FOX, R. H. "On Topologies for Function Spaces." *Bull. Amer. Math. Soc.*, **51** (1945), 429–32.

FRINK, A. H. "Distance Functions and the Metrization Problem." *Bull. Amer. Math. Soc.*, **43** (1937), 133–42.

HANNER, O. "Retraction and Extension of Mappings of Metric and Non-metric Spaces." *Ark. Math.*, **2** (1952), 315–60.

HEATH, R. W. "Screenability, Pointwise Paracompactness, and Metrization of Moore Spaces." *Can. J. Math.*, **16** (1964), 763–70.

JONES, F. B. "Concerning Normal and Completely Normal Spaces." *Bull. Amer. Math. Soc.*, **43** (1937), 671–77.

———. "Metrization." *Amer. Math. Monthly*, **73** (1966), 571–76.

———. "Remarks on the Normal Moore Space Metrization Problem." *Proc. of Wisconsin Summer Topology Seminar (1965)*, *Ann. Math. Studies*, **60** (1966).

KELLEY, J. L. "Convergence in Topology." *Duke Math. J.*, **17** (1950), 277–83.

———. "The Tychonoff Product Theorem Implies the Axiom of Choice." *Fund. Math.*, **37** (1950), 75–76.

KNASTER, B., and C. KURATOWSKI. "A Connected and Connected *Im Kleinen* Point Set Which Contains No Perfect Set." *Bull. Amer. Math. Soc.*, **33** (1927), 106–109.

KNIGHT, C. J. "Box Topologies." *Quart. J. Math. Oxford*, **15** (1964), 41–54.

LEUSCHEN, J. E., and B. T. SIMS. "Stronger Forms of Connectivity." *Rend. Circ. Math. Palermo*, **21** (1972), 255–66.

LEVINE, N. L. "A Characterization of Compact Metric Spaces." *Amer. Math. Monthly*, **73** (1961), 657–58.

MANSFIELD, M. J. "Some Generalizations of Full Normality." *Trans. Amer. Math. Soc.*, **86** (1957), 489–505.

MICHAEL, E. "A Note on Paracompact Spaces." *Proc. Amer. Math. Soc.*, **4** (1953), 831–38.

———. "Another Note on Paracompact Spaces." *Proc. Amer. Math. Soc.*, **8** (1957), 822–28.

———. "The Product of a Normal Space and a Metric Space Need Not Be Normal." *Bull. Amer. Math. Soc.*, **69** (1963), 375–76.

———. "Yet Another Note on Paracompact Spaces." *Proc. Amer. Math. Soc.*, **10** (1959), 309–14.

MOORE, E. H., and H. L. SMITH. "A General Theory of Limits." *Amer. J. Math.*, **44** (1922), 102–21.

MOORE, R. L. "A Connected and Regular Point Set Which Contains No Arc." *Bull. Amer. Math. Soc.*, **32** (1926), 331–32.

———. "Concerning Connectedness *Im Kleinen* and a Related Property." *Fund. Math.*, **3** (1922), 232–37.

———. "Concerning the Cut Points of Continuous Curves and of Other Closed and Continuous Point Sets." *Proc. Nat. Acad. Sci.*, **9** (1923), 101–106.

MORITA, K. "On the Product of a Normal Space with a Metric Space." *Proc. Jap. Acad.*, **39** (1963), 148–50.

———. "On the Product of Paracompact Spaces." *Proc. Jap. Acad.*, **39** (1963), 559–63.

———. "Star-Finite Coverings and the Star-Finite Property." *Math. Jap.*, **1** (1948), 60–68.

NAGATA, J. "On a Necessary and Sufficient Condition of Metrizability." *J. Inst. Polytech. Osaka City Univ.*, **1** (1950), 93–100.

NOVAK, J. "On the Cartesian Product of Two Compact Spaces." *Fund. Math.*, **40** (1953), 106–12.

PERVIN, W. J. "On Separation and Proximity Spaces." *Amer. Math. Monthly*, **71** (1964), 158–61.

RIBEIRO, H. "Sur les espaces à métrique faible." *Port. Math.*, **4** (1943), 21–40, 65–68.

ROSS, K. A., and A. H. STONE. "Products of Separable Spaces." *Amer. Math. Monthly*, **71** (1964), 398–403.

RUDIN, M. E. "A Separable Normal, Non-paracompact Space." *Proc. Amer. Math. Soc.*, **7** (1956), 940–41.

SANDERSON, D. E., and S. K. HILDEBRAND. "Connectivity Functions and Retracts." *Fund. Math.*, **57** (1965), 237–45.

———, and B. T. SIMS. "A Characterization and Generalization of Semi-metrizability." *Amer. Math. Monthly*, **73** (1966), 361–65.

SCARBOROUGH, C. T., and A. H. STONE. "Products of Nearly Compact Spaces." *Trans. Amer. Math. Soc.*, **124** (1966), 131–47.

SIMS, B. T. "Between T_2 and T_3." *Math. Mag.*, **40** (1967), 25–26.

———. "Some Generalizations of the Contraction Mapping Theorem." *Rend. Circ. Math. Palermo*, **21** (1972), 64–70.

SION, M., and G. ZELMER. "On Quasi-metrizability." *Can. J. Math.*, **19** (1967), 1243–1249.

SMIRNOV, Y. M. "A Necessary and Sufficient Condition for Metrizability of a Topological Space." *Dokl. Akad. Nauk SSSR*, **77** (1951), 197–200.

———. "On Metrization of Topological Spaces." *Amer. Math. Soc. Transl.*, **91** (1953).

SORGENFREY, R. H. "On the Topological Product of Paracompact Spaces." *Bull. Amer. Math. Soc.*, **53** (1947), 631–32.

STEINER, A. K. "The Lattice of Topologies: Structure and Complementation." *Trans. Amer. Math. Soc.*, **122** (1966), 379–98.

STONE, A. H. "Paracompactness and Product Spaces." *Bull. Amer. Math. Soc.*, **54** (1948), 977–82.

TAMANO, H. "On Paracompactness." *Pac. J. Math.*, **10** (1960), 1043–1047.

TYCHONOFF, A. "Über einen Metrisationsatz von P. Urysohn." *Math. Ann.*, **95** (1926), 139–42.

URYSOHN, P. "Über Metrization der Kompakten Topologischen Räume." *Math. Ann.*, **92** (1924), 275–93.

———. "Zum Metrisationproblem." *Math. Ann.*, **94** (1925), 309–15.

VAN ROOIJ, A. C. M. "The Lattices of All Topologies Is Complemented." *Can. J. Math.*, **20** (1968), 805–807.

WALLACE, A. D. "Separation Spaces." *Ann. Math.*, **42** (1941), 687–97.

WALLMAN, H. "Lattices and Topological Spaces." *Ann. Math.*, **39** (1938), 112–26.

WHYBURN, G. T. "On the Structure of Connected and Connected *Im Kleinen* Point Sets." *Trans. Amer. Math. Soc.*, **32** (1930), 926–43.

———, "Open and Closed Mappings." *Duke Math. J.*, **17** (1950), 69–74.

YOUNG, G. S. "The Introduction of Local Connectivity by Change of Topology." *Amer. J. Math.*, **68** (1946), 479–94.

Index

D

E

F

G

H

I

J

K

L

M

N

O